THE AMATEUR NATURALIST

THE AMATEUR NATURALIST

Explorations and Investigations

by Charles E. Roth

An Amateur Science Series Book
FRANKLIN WATTS
New York · Chicago · London · Toronto · Sydney

For Bryan,
Jessica, and
Sara Jane,
amateur naturalists
of the future

Illustrations by Glenn Wolff

Photographs copyright © :
Photo Researchers Inc.: p. 39 (Robert J. Erwin/NAS);
Visuals Unlimited, Inc.: pp. 70 (Glenn M. Oliver),
78 (John D. Cunningham).

Library of Congress Cataloging-in-Publication Data

Roth, Charles Edmund, 1934–
The amateur naturalist : explorations and investigations / by Charles E. Roth.
p. cm. — (An Amateur science series book)
Includes bibliographical references (p.) and index.
Summary: Presents a variety of activities and projects to help the amateur naturalist explore the habits of common plants and animals.
ISBN 0-531-11002-8
1. Natural history projects—Juvenile literature. 2. Nature study—Juvenile literature. [1. Natural history projects. 2. Science projects.] I. Title. II. Series.
QH55.R67 1993
574'.078—DC20 93-13390 CIP AC

Printed in the United States of America
6 5 4 3 2 1

Contents

THE AMATEUR NATURALIST

Who Shares Your Neighborhood?

Do you know who your nonhuman neighbors are? Do you know where they live and what they do with their lives? No one knows all of these neighbors and most know the details about the lives of very few. The people who know the most about such things are naturalists.

If you are curious about your wild neighbors, one of the first things you can do is begin to make a list of who they are and where they live near you. You may have heard about people who keep a "little black book" with the names and addresses of those with whom they want to keep contact. A first project for budding naturalists is to develop such a "little black book" of the plants and animals that live near you.

With some careful looking you will find nonhuman neighbors almost anywhere you look. There are spiders, insects, and mice that may well spend time in your home from time to time and there are a variety of animals and plants that share yards and nearby vacant lots as well as woodlands and waterways. Many are small and inconspicuous; a few are large and obvious. Most have interesting ways of making their living in this world when you get to know them, and a few are truly amazing.

Like human friends, some will be more interesting to you than others. But to make that judgment you must first make at least an initial acquaintance.

PROJECT: DEVELOPING A LITTLE BLACK BOOK OF WILD NEIGHBORS

Any small notebook will do. A spiral-bound, lined notebook is probably easiest to begin with. Some people I know like the small black, bound, unlined sketchbooks many artists use. That is because they like to sketch portraits of the wild neighbors they meet as well as write descriptive words about them.

In setting up your book it is useful to set aside sections for different kinds of neighbors—a section for birds, for mammals, for insects, for spiders, for trees, for small plants, and the like. At first you may have more entries in one section than in the others because you are initially looking for those things. As you investigate the area, you are almost certain to spot other things that interest you and you should note them as well.

When you first spot a wild neighbor that interests you, you may very well not know its name. In that case just give it a number. Then after the number write down a description of what it looks like—its size, shape, color, movements, sounds. Write down or make sketches of anything and everything about it that would help you recognize it again.

You will also want to note its address, that is, where you saw it. If it is a plant, you can probably come back often and find it in that place at least for a growing season. If it's an animal, you may or may not find it in exactly the same place again. It is likely to be found in the same general area but not the very same place.

Leave a space in each entry to add the name of the creature once you learn it. The descriptions you wrote or sketched will provide many of the clues you will need to track down a name for it. You can use the descriptions to look in field guides written by skilled field naturalists or to ask knowledgeable people who you think might know what it is.

Entries in the "black book" might look something like the following samples:

#27—this plant is growing along the roadside about two blocks from home. It has leaves like a dandelion but it is much taller. The flowers are bright blue. The flowers are open in the morning when I go by but by late afternoon they are closed.

Underneath the bird feeder this morning I saw a gray mouselike critter with a very pointy nose. It had a short, pointed tail, too, which made it seem hard to tell whether it was coming or going. It seemed very quick and nervous. It would pop out from a hole in the snow and grab a sunflower seed that had fallen from the feeder and quickly dive back into the snow hole.

#35 Down in the ivy that grows on the church there is a pair of interesting birds. The bright-colored one, probably the male, looks as if it has been dipped in raspberry juice. It has a pretty warbling song. Its mate is stripy and sparrow-like. The sparrowlike one keeps flying back and forth with sticks and grasses, and I think it is making a nest in the ivy. I haven't seen the colored one with nesting material. It does stay near the ivy and sings a lot.

Went swimming today and in the sandy shallow area watched some fish making saucerlike places in the lake bottom. A fish would pick up things in its mouth from the spot, swim a little way, and spit it out. Then it would come back to the spot and swim around, and its motion would brush other junk away. If another fish came near the spot it would dash right at it and drive it away. I'm pretty sure the fish are some kind of sunfish but I don't know just what kind. I'm going to have to come back there from time to time and find out what they do with those cleared places.

Heard thumping on the window screen and investigated. Found a very large pale green moth with long "tails" on its hind wings. Its antennas were large and like little ferns attached to its head. I had heard about these but this was my very first luna moth. I am told they were very rare for many years, when pesticide spraying over large areas was common. Now they are making something of a comeback. How I wanted to catch it! However, I let it stay free. Perhaps I will be able to find luna caterpillars on the hickory trees in the woodlot down the street later this summer.

(For the solutions to these "nature detective" stories, see the end of this chapter.)

Wherever you live in this country, there are interesting plants and animals whose lives you can investigate. The hallmark of a naturalist is curiosity about the other living things that share your surroundings and the special worlds in which they live. As a naturalist you have the desire to investigate and develop the skills and patience to make observations over time. You will then devise ways to ask questions of at least some of your wild neighbors that can be answered by observing them and watching what they do in nature or under special conditions that you create.

Being an amateur is something to take pride in. It means you are doing something for the love of it whether or not you get paid for doing it. Some fine naturalists earn their living in quite different fields and of course there are some professional naturalists who qualify as amateurs because of their attitude toward their work; they feel very fortunate to be paid to do those things they most love to do.

Making entries in such a book is only a beginning project for a beginning naturalist. While you are looking about for creatures to enter in the book you will undoubtedly find that some of them catch your interest more than others. Just why one appears to be of greater interest than another differs from person to person. Actually, why is of less importance than the fact that some do. You should circle those that interest you the most. These are the ones you will want to visit repeatedly and make more observations and studies of. These are the ones you will probably work hardest to find a name for. This book is designed to provide you with activities and techniques to dig in and discover more and more about the lives of those wild neighbors that most intrigue you.

Plants and animals described in "black book" entries: #27: chicory; short-tailed shrew; #35: house finch; bluegills.

Asking Questions of Nature

You will find that some of your wild neighbors seem to be more interesting or intriguing to you than others. You will have listed more questions about them than many of the others—they may be birds, or insects, or plants, it doesn't really matter. The challenge is to find ways to pose questions about them that they can help answer for you. Obviously you can't ask them in your language, you must find ways to get them to provide answers by their behavior. That is language that you can learn to understand.

In literature, Dr. Doolittle was supposed to be able to talk to the animals. You won't be able to do it as he was supposed to, but you can set up ways to get them to answer many of your questions. For example:

You might want to ask a male red-winged blackbird how big his territory is. You could get answers from him by setting up the following conditions . . .

PROJECT: INVESTIGATING REDWING BLACKBIRD TERRITORIES

Redwing blackbird males display to other males whenever another male enters the space a particular male has claimed for his own. The display involves dropping the shoulders and showing the bright red wing patch to the invading male (Figure 2-1). Redwings cannot tell their

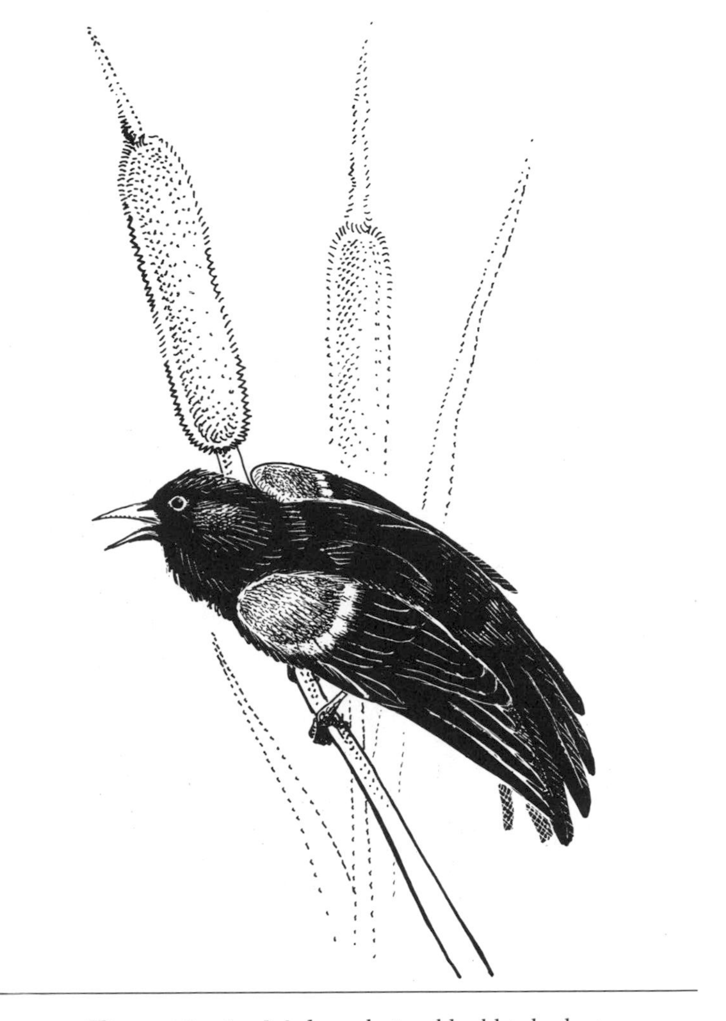

Figure 2–1. *Male redwing blackbirds drop their shoulders and show a bright red patch to other males who "invade" their territory.*

reflection in a mirror from a real bird. A male regularly patrols its territorial boundaries. If it finds a mirror it will display to its own reflection.

Mount a mirror on a pointed stake that is about 5 feet tall (1.5 m) (Figure 2-2). Place the mirror at a point you think is in the redwing's territory. Move away and watch through binoculars to see if the bird finds the mirror and displays at it. On a map you make of the

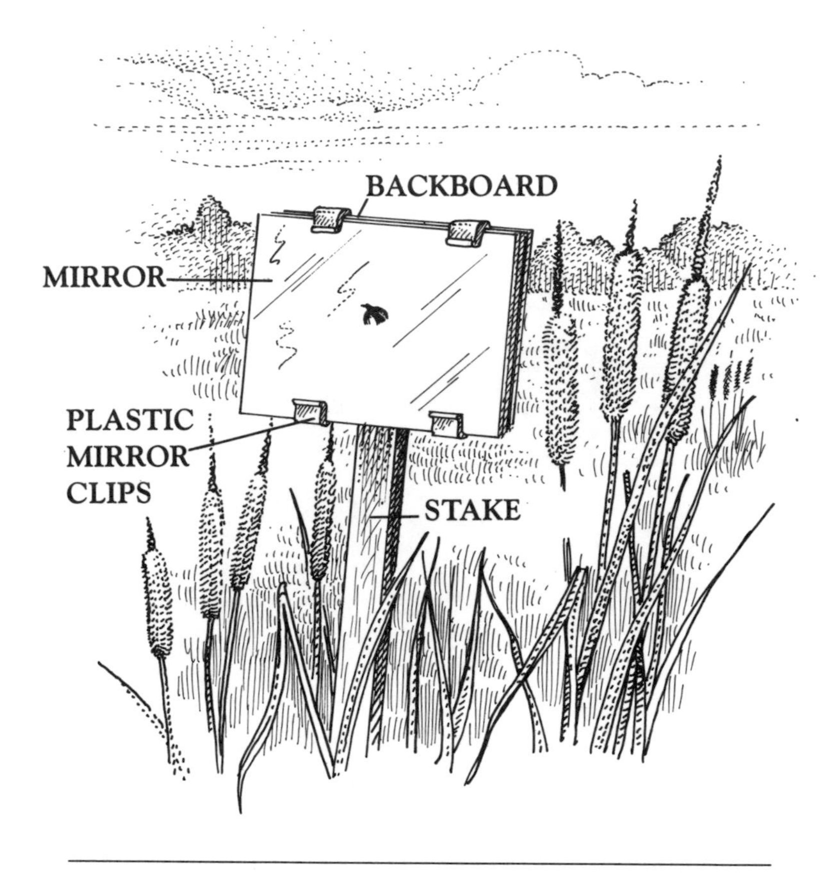

Figure 2–2. *You can investigate redwing blackbird territoriality by mounting a mirror on a pointed stake. The males can thus be tricked into displaying to their own reflection in the mirror.*

marshy area where the bird is found make a mark to indicate where you stuck the stake in the ground. Move the stake and mirror around and repeat your observations. On the map, circle each place where the bird displayed to the mirror. Make a line connecting all the circled points and you will have a rough outline of that bird's territory.

You will have to decide how far to move the stake each time. If the distance is too great you will not get enough display activity to make a good map. If your moving distance is too short, you will get a lot of information but you are likely to disturb the bird too much. Try moving the stake 3 to 5 feet (.9–1.5 m) each time for a start.

The shape of a particular bird's territory will vary. Some will be nearly circular; others will be oblong.

Begin this project in early spring, shortly after the males arrive from the south. They set up their territories before the females arrive. The females choose the territories they feel have the best nest sites. The male that has staked out the territory chosen by a particular female will become her mate.

Once the females have arrived and have begun to build a nest, stop this project. Your movement around the territory might disturb the nesting. Such movement may also attract predators that might find and destroy the eggs or young.

Other birds also defend territories and will also challenge their reflections in a mirror. Many males regularly sing from specific perches. First determine where a male is to be found singing regularly. Set up the mirror nearby and make your observations. Map all the places where you place the mirror. Connect the places where the bird displays in order to get an idea of the territory it calls home. In addition to robins, try investigating cardinals or mockingbirds. These birds often attack their reflections in windows or car hubcaps that are found in their territories.

Or you might wonder if wooly bear caterpillars wander at random during the fall or if they prefer to go in a particular direction. You could get them to tell you by . . .

PROJECT: INVESTIGATING WOOLY BEAR WANDERING

Collect a group of wooly bear caterpillars (Figures 2-3 and 2-4) and put them in a covered container. Find an open area of dirt or even paving on which you can draw a circle about 6 to 8 feet (1.8–2.4 m) in diameter. In the center of the big circle make a smaller circle about a foot (.3 m) in diameter. Use a compass to mark out the major points of north, south, east, and west on both circles. On a piece of paper make a drawing of your two circles.

Once the circles have been laid out on the ground and on paper, release the caterpillars into the center circle. Keep track of each of the caterpillars and map its route until it reaches the boundary of the big circle. It can then be recollected and returned to its jar for further questioning.

If you have more than four or five caterpillars you may want to get some friends to help you map the routes of some of the caterpillars.

What kind of a pattern do the paths of the caterpillars make? Do they go in all different directions or do they tend to go more in one direction than another? If you have the caterpillars repeat the task, is the pattern the same or different?

Figure 2–3. *Wooly bear caterpillars are the young of the Isabella moth.*

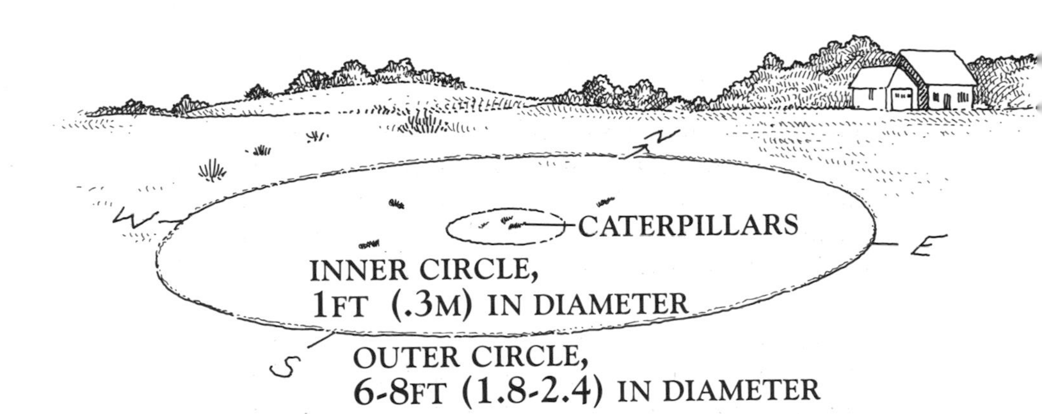

Figure 2–4. *You can investigate the movements of wooly bear caterpillars. Keep track of the path of each of the caterpillars and map its route.*

Do the same study on different kinds of days: sunny, overcast, partly sunny. Do the paths of the caterpillars show the same or different patterns? Does it make a difference if the caterpillars are put out into the circles in the morning or afternoon?

Keep track of the path of the sun while you are asking your questions of the caterpillars. Also keep track of the temperature. Be sure to take temperature readings at caterpillar level rather than at your standing level. Does there seem to be any relationship between the position of the sun and the routes the caterpillars travel? Does their degree of activity seem to be affected by temperature?

The rusty brown and black-banded wooly bear caterpillars are the young of a tiger moth known as the isabella moth. The caterpillars feed on a variety of plants but commonly on the leaves of plantain, a common weed of lawns and fields. They often aren't seen until they are ready to seek a spot in which to spend the winter. It is then that they become active wanderers, and they eat little if anything at that time. They spend the winter as caterpillars under rubbish. In spring they come out and feed for several weeks until they are ready to make a cocoon of silk and caterpillar hair. In this they pupate. They will emerge as adult moths in about two weeks. The cocoons are commonly found under boards and stones.

Folklore has it that one can predict the severity of the coming winter by the width of the brown bands on the caterpillars. Actually the brown bands seem to get broader as the caterpillars get older and larger. People who ask the caterpillar for information on the coming winter by measuring the band width of large numbers of them and comparing that with the winter weather do not come up with any good or consistent answers from the caterpillars.

Black, as a color, absorbs heat better than any other color. Do you think that the location of the black bands at the front and back of the animal helps explain any of the patterns of movements you observed? Do you think this relates at all to the location, presence, or absence of the sun?

You might ask an ant lion why it always seems to make its pits in places protected from the rain. If you do the following things the ant lions will probably reveal their secrets to you . . .

PROJECT: INVESTIGATING ANT LION DIGGINGS

You may have noticed in patches of dry sandy soil small cone-shaped pits. You may have wondered what caused them. If you take a spoon or trowel and quickly dig under a pit and then spread the material out on a piece of white paper you will likely discover a weird little bug among the grains of sand or you may find a perfect little ball of sand grains stuck together. The little flattish, brownish bug with big pincers, small eyes, and covered with lots of small hairs is called an ant lion. It won't bite you, the big pincers are for grabbing ants, which are its primary food. It will eat other small insects, too. The ball of sand grains you may find instead of the insect is the cocoon that protects the insect in its pupal stage as it goes through its change to an adult insect. Usually where you find one ant lion pit you will find several more, often a whole colony of them.

Adult ant lions are active at night. They don't look a bit like the ant-eating larvae. Instead they look very much like damsel flies. They have a long, thin body with a small head, large eyes, and lacy wings. You can tell them from damsel flies or dragonflies because the ant lion adults have fairly long antennas, usually clubbed. Dragonflies and damsel flies have only very short, straight antennas.

There are a lot of questions you might wish to ask of these strange little larvae such as:

- How do they move?
- Why do they make the pits?
- How do they make the pits?
- What kind of ground or material do they need in order to make proper pits?
- What kind of things do ant lions react to?
- How quickly can ant lions find a new place to make a pit?
- What kind of barriers can an ant lion cross?
- How many ants can an ant lion eat in a day?
- Do ant lions that eat more grow up faster than those that eat less?
- Do ant lions prefer wet or dry sand?
- Do ant lions prefer sunlight or shade?
- Do ant lions move around very much?

You can probably think of even more questions. With some creative thinking you can create situations where the ant lions themselves will answer each of your questions.

Any box that is about two inches deep and can hold

sand or similar material is suitable for asking questions of ant lions (Figure 2-5). You might begin by putting just a thin layer of sand on the bottom of the container. The layer should be too thin for the insects to bury themselves in. Place some ant lions in the box. Observe how they move. They will soon have given you the answers to the first question on the list above.

Next you can add more sand, about an inch, to the container. Watch and record what the ant lions do. Be patient. They may be shy and try and just hide at first because you have handled and frightened them. You may even have to leave them alone for a while and then come back to observe.

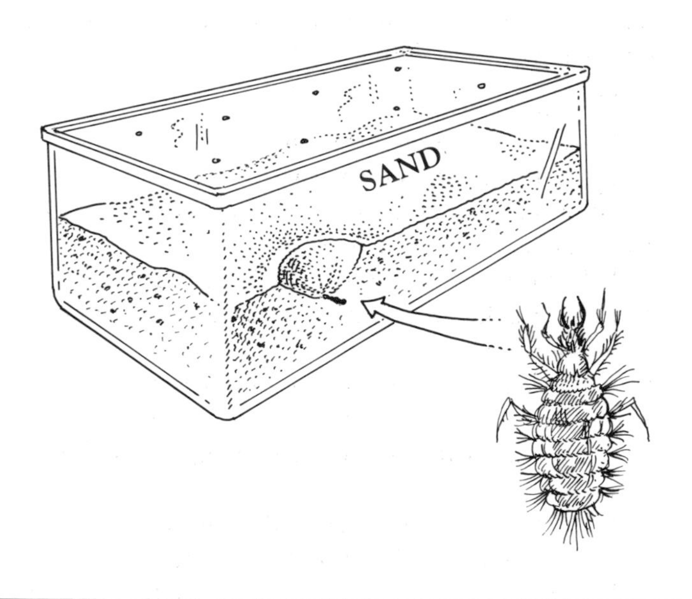

Figure 2–5. *Ant lion larvae can be found in the cone-shaped pits often seen in sandy soil. Fill a box with sand to recreate its natural habitat and study its activities.*

Once some pits have been made, you can add some ants to the container. You will then want to put a lid on it so the ants cannot climb out. This may be clear plastic so you can observe what happens. If you don't feed the ant lions they will die after a few days. How do the ant lions catch the ants?

Use your imagination and devise things to do in the container to get the ant lion to reveal the answers to your questions. Once the captive ant lions have given you an answer, you may want to set up a similar experiment with wild ant lions to see if they give essentially the same answer. If you are very careful, you can put a tiny spot of nail polish on individual ant lions so that they can be recognized from the others. Nail polish in larger amounts can poison the creatures, so keep the spot very tiny.

PROJECT: INVESTIGATING BIRDS' FEEDING PREFERENCES

Do the different birds you have coming to your backyard feeder like certain seeds better than others? You can set up a feeding tray with partitions that hold different seeds, not only those in purchased mixes but also seed that you have collected in the neighborhood (Figure 2-6). Observations of which birds choose what seeds will give you some answers. You may also discover that their preferences change during the feeding season.

The challenge always is to create situations where the plant or animal you are interested in can answer your questions by what they do or do not do. This will challenge both your imagination and your powers of observation. If you do both well, the creatures will give you reliable answers again and again. In fact, other observers, doing the same things, will get very similar, if not the same, answers.

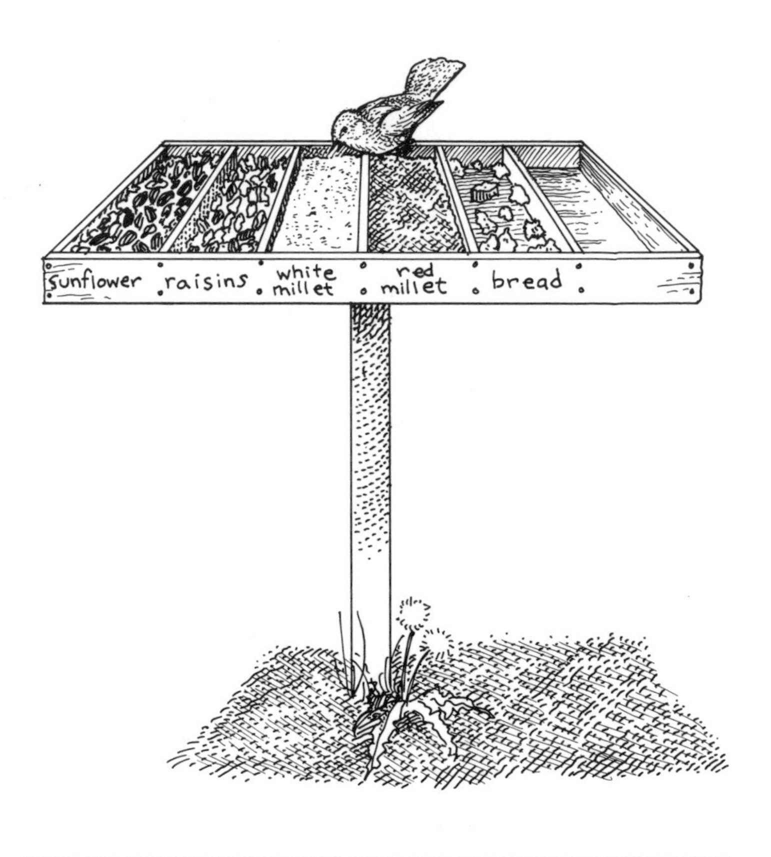

Figure 2–6. *A feeding tray will help you to investigate what kinds of seeds and other foods different types of birds prefer.*

You can invent the situations and make the observations by yourself but it is often easier and more rewarding to do it with others. Sharing thoughts and observations often results in more creative setups and more reliable observations.

Looking and Listening for Answers

Choosing the questions you want to ask of the natural world and designing ways to get nature to give you some answers are only the first steps to becoming an amateur naturalist. You have to develop the patience to truly observe, that is to look and listen, for the answers as they appear. You will also have to develop skills at recording your observations. Being an amateur naturalist is a lot like being a detective; you have to record as many clues as possible. Hidden among these clues often will be the answers to your questions. However, you may have to look at the clues repeatedly before you find the key ones.

The first step is to write down what you see, hear, and smell. You can do this with words and drawings (Figure 3-1). It doesn't have to be with perfect grammar or spelling. It should, however, make sense to you when you read it even weeks or months later.

Some things should be noted every time you start taking notes, such as:

- date
- time of day
- weather conditions
    - —temperature
    - —amount of cloud cover
    - —wet or dry conditions

Figure 3–1. *A sample nature journal entry*

The reason for doing this is that these factors often affect the way plants and animals behave: Get in the habit of recording this information. Then you will always have it if you need it. You may not always need it to make sense of your observations. On the other hand, if you later suspect that the season and the weather may be affecting what the creature you are investigating does, you are lost without having recorded the information.

You can make the task easier by using some symbols to help record the data:

Record date in boxes or with slashes: Month/day/year (10/16/91).

Time: Use circle with dot in center. Draw in clock hands, adding A.M. or P.M.

Cloud cover: Use circle and fill in percentage of sky covered by clouds.

Temperature: Use thermometer symbols. It's good to use at least two, recording on one temperature at your eye level and one at the animal's eye level or the height of a plant or its flower.

Wet/Dry: Use letters D for dry; H for humid; W for wet or raining. Circle the most appropriate letter.

If you keep a naturalist's journal, you can put these symbols at the top of each page. Use a new page for each observation. It's fun during bad weather or evenings to read through your journal and remember different observations you have made. Search for patterns in your observations that reveal unsuspected information about some of the creatures you have observed over time.

Activity that you have recorded again and again tells you that such action is normal for that creature. Other activity you will see less often and only under certain conditions. The fun is in trying to determine just what the conditions are under which that activity occurs. For example, flying squirrels are normally active after dark. My notes show that I have sometimes seen them out during the day. Further study of those notes shows that the days I have seen them active have always been days with a heavy overcast. On the few other occasions that I have seen them out in bright light, it was only because they had come out of their holes after I repeatedly knocked on hollow trees with woodpecker holes in them. The squirrels had been sleeping in those old woodpecker holes. They quickly find new places to hide until dark.

Your journal notes will most often be written in words and some shorthand symbols that you use, such as arrows to show directions of motions. They will also probably include a variety of sketches of objects. You can record much information through sketches and diagrams. They don't have to be great art. They just have to remind you of key shapes and patterns. They might include such things as a quick sketch of a bird to note where there were particular colors or patterns. It might be a tracing of the outline of a leaf with the vein pattern sketched in. Such things will help you look up and identify in reference books something you saw in the field but couldn't immediately recognize.

You don't have to limit what you note in the journal to observations. Record your questions, your thoughts about your observations, your feelings about the experience, and any hypotheses that may be triggered about the how and why of what you have seen. Such thoughts and feelings are slippery. Unless you jot them down promptly they are apt to disappear. Those very thoughts reopen useful patterns of thinking many times when you reread the journal.

If the thought of doing much writing makes you feel uncomfortable, you can create data sheets that you can quickly fill in with check marks or single words. Making up such sheets takes a bit of planning before you set about doing your investigation. You must first decide what kind of information you need to collect.

Let's say you want to find out the kinds of seeds preferred by the variety of birds that visit a bird feeder. You will need a list of the kinds of birds that usually come to the feeder. You will also want a list of the kinds of seeds, or other foods, that you put out on the feeder.

First make a chart of the birds. It may or may not include a picture of the bird to help you or any other observers identify the visitors. Assign each bird a letter symbol, such as C for chickadee, N for nuthatch, CD for cardinal, BJ for blue jay, and so forth.

Next, make a seed visitation chart. It may look similar to the one below. Each time a bird visits the feeder and takes a seed, jot down its letter symbol on the lines following the type of seed it chose.

Date: ____________ Observation time: Began ____________ Ended ____________	
Weather conditions	
Seed Type	Birds
black sunflower	
striped sunflower	
millet	
rape	
cracked corn	
oat cereal	

It will of course be easier to make your observations accurate if you divide the feeding tray into sections. Put one kind of seed in each section. If you mix the food on the tray or feeder it is more difficult to tell what seed a particular bird has selected.

Once you have determined a particular kind of bird's favorite food you might ask if it favors one color over another. You could do this by putting colored papers on different parts of a tray or other feeder. Place the favorite food on each of the different colors. You would then set up a similar record sheet.

Colors	Number of visitors (name bird species studied)
blue	
red	
yellow	
green	
white	
black	
brown	

All the observer has to do is to make a check after the proper color each time a bird of the kind you are studying comes to that color. If you fill out the chart neatly you will have created a sort of bar graph that will show a pattern of birds' preferences. You can of course add up all the numbers. Use those numbers to develop some other types of graphs that will show patterns among the data.

Games to Help Sharpen Observation Skills • There are several exercises that you can do to improve both your observation skills and your perceptions of patterns among the things observed. Most of these exercises are best done with a partner or a small group of friends.

Rudyard Kipling wrote a novel named *Kim.* In the novel some of the street kids are trained to be thieves by an evil adult thief. One of their activities has become known as Kim's game, and it was used to train the boys to be sharp observers. To play you will need a shallow box and a cloth with which to cover it. Begin by placing four or five objects in the box (more can be added later as you become more skilled).

One person arranges the objects in the box without the others watching and then covers the box. Give the others the chance to view the uncovered box for a very brief time. Start with about thirty seconds and reduce the viewing time as the participants become more skilled.

The person in charge snaps off the cover for the allotted time and then recovers it. The players then have to write down as complete a description as possible of what they saw in the box—the type of items, the number, their placement, and the like.

When you have several players it is interesting to compare notes and see what different items each player saw and recorded. Different people see the same item through different "filters" of background and experience. It's the same way in making natural history observations. Two people will see the same event but record and interpret it differently. It often takes many observations of the same event before widespread agreement on what actually happened is reached among people.

Repeat the procedure of Kim's game over and over. Each time change the number and type of placement of objects in the box. Rotate the roles of the "person in charge" and the "players." It's a fun party game as well as excellent training for the serious young naturalist.

A different version of this game can be played by two people when they are out in the woods or in a field or even in a vacant lot. The two people sit on the ground facing each other within arm's length of each other. They study the ground before them carefully and familiarize themselves with all they see.

One then closes his or her eyes while the other rearranges something in the patch of ground they have been examining. That person then announces "open eyes" and records how long it takes for the other to spot what has been changed. Then the roles are reversed. This can be repeated over and over to see just how slight a change the other can detect.

Many animals have developed the ability to spot slight changes in their surroundings to a very high degree. They know what is normal in their home grounds. They are ever alert to the smallest changes—a new object, something out of place, an unusual motion, a new smell. Thus, they spot even the observer that tries to camouflage himself or the blind in which an observer is hiding. Such blinds often have to be put in place days or weeks before the observer intends to use them so that the animals come to accept it as a new item in their world.

Another skill-building activity is to walk casually through an unfamiliar home, office, park, or other place that you can revisit. After going through, jot down everything that you can remember seeing. Now go back and look very carefully and compare it with the list you made. What did you miss the first time through? Were there whole groups of objects that you missed seeing? This can be done alone or with a friend. If you do it with a friend, compare your lists. Do you each see the same items or do each of you tend to focus in on or miss different kinds of items? Repeat this activity in different places until you become better and better at getting a really full picture the first time through.

These activities for both improving your observations and finding patterns in what you see will help make you a better naturalist.

Knowing What Creature You Have Observed • It seems to be a distinctly human trait to find a special "pigeonhole" for everything. We want to have a place for everything and everything in its place. From the earliest time people have tried to classify everything. One very early classification was animal, vegetable, and mineral. We still quickly use such classifications today when playing guessing games such as twenty questions.

Unfortunately there are so many different things on this planet that putting them all into only three boxes creates groups far too large to be useful. Early naturalists devoted much time to creating much smaller divisions that would help us all to better understand the different things being observed. John Ray, who lived between 1627 and 1705, created the concept of species; that is, all beings that have essentially the same range of appearance and behavior. The offspring of two members of the same species are essentially the same as their parents. Offspring of two different species may carry appearances of both parent species. In general, two different species cannot successfully mate and have offspring, or if they do, these offspring cannot have offspring themselves.

Based on the concept of species, many early naturalists came up with groupings of species based on supposed similarities and differences. It was Swedish naturalist Carl von Linné who devised a scheme for classification that became widely accepted and, with modification, is in use today. Carl was an active observer of plants from his earliest days. At eight years old he was nicknamed "the little botanist" by friends and relatives. He went on to make the study of plants his life's work. By the mid-1730s he had pulled his thoughts on classification together in a book called *Systema Naturae.* He went on to produce several important books on plants that developed more fully his ideas about classification. Two of these, *Species Plantarum* and the fifth edition of *Genera Plantarum* are considered to be the official starting points of modern classification of living things. These were published in 1753 and 1754.

The titles seem strange to us today for they are in Latin. However, the heart of the system he developed was to give each species a two-word name, much like our modern surnames and given names. The first name of a species tells what larger group, or genus, the species belongs to, similar to the way our surname tells who our family is. The second name is the species name. Latin is still used for scientific names because it is considered a "dead" language that changes very slowly, unlike modern "living" languages. The Latin names are recognized by scholars in nations around the world. Carl von Linné himself changed his name by Latinizing it. Today he is known as Carolus Linnaeus.

As the science of classification grew, more and more effort was made to use the classification system to show relationships between different groups of organisms. Species were grouped together into a genus based on apparent relationships between the species. Groups of similar genera (plural of genus) were known as families. Families with some similarities were grouped together as a phylum (plural phyla). Groups of similar phyla were considered a kingdom. Thus we can list the groups from broadest to most specific as follows:

Kingdom
 Phylum
  Family
   Genus
    Species

There are even finer divisions used by scientists who classify living things, but they need not concern us here.

As more and more people have explored the world of living things they have discovered more and more ways to determine relationships between living things and even how long ago one species branched off from its relatives to start its own line of development. The ongoing study of these relationships means that you may find different classifications for a given species in different sources. In fact there is a very specific set of rules among scientists for changing a species' classification. Sometimes it is confusing to laypeople.

Even the largest divisions have been undergoing change. Only a relatively few years ago people believed there were only three kingdoms of living things—Plants, Animals, and Protists, the microscopic creatures. Today five kingdoms are generally recognized and some scientists believe there may actually be several more. This has emerged as we have learned more about the amazing diversity of microscopic and submicroscopic creatures. The accepted five kingdoms now include Plant, Animal, Fungi, Protoctista (largely microscopic creatures that have a nucleus in their cells), and Monera (microscopic creatures that lack a cell nucleus).

When you ask the question, "What kind of living things am I observing?" you will normally turn to field guides and technical keys for the answers. Of course you can ask an expert but they will have used such guides and keys somewhere along the line. These guides and keys are built around the modern system of classification that has been developed from Linnaeus's work two centuries ago. There are scientists and naturalists in museums and universities around the world who continue to explore the questions about how one species is related to another and how the other classification groups are also related. If you are truly puzzled about a creature and cannot find it in any guide or key, you can contact such places and track down the experts who may be able to help you answer your questions.

Asking Questions of Animals

Almost anywhere you live there is some form of animal life, and you can figure out ways to get them to reveal secrets about their lives to you. Of course, large mammals and birds may not be common where you live but there is a great host of other animal life to challenge your curiosity. Remember that animals include fish and reptiles, insects and spiders, snails and worms, and many other creatures that are not plant or fungi.

Some of your questions about any of these creatures may be answered by simple observations of neighborhood animals. Other questions may require simple experiments with animals held *temporarily* captive.

Naturalist Henry Beston wrote "Animals are not brethren, they are not underlings; they are other nations, caught with ourselves in the net of life and time." We must always be aware of that as we interact with other creatures. Temporary captives must be treated with all kindness and concern for their health and well-being. They should be returned to the places they originally came from as soon as possible. Apply the golden rule of "doing unto others as you would be done to" to the wildlife around you as well as to other people.

PROJECT: KEEPING TRACK OF HOW ANIMALS USE THEIR TIME

You know that you live a busy life. What kind of life do your wild neighbors live? You get up in the morning, eat, bathe, dress, do various activities, socialize, sleep, and the like. There is usually some sort of pattern to how and when you do these things. Is the same true of your wild neighbors?

You can get answers to these questions in part by taking the time to closely observe animals around you. Members of the squirrel family are good animals to do this with. Most are active during the day as we are and they adjust readily to being watched.

Begin by gathering information about a few of the more obvious activities. Later you can add more to your list as you gain experience with what they do with their lives. Watch the squirrels whenever you can (Figure 4-1). Use a checklist like the one on page 38. Write in the number of minutes the animal spends on each activity.

Once you get an idea of how much time is spent on such activities you may want to ask some more questions and hunt for the answers using similar information-gathering charts. For example, are the squirrels more likely to be foraging at certain times of the day than others? To answer such a question you need to record not only how many minutes they spend foraging but the exact times of day you see them foraging. Instead of noting

Foraging: 10 minutes

you would note

Foraging: 7:45–7:50
Eating: 7:50–7:55
Foraging: 7:55–8:12

Date________	Observation Place:__________				
Activity	Time Spent on Activity				
	Animals Observed				
	#1	#2	#3	#4	#5
resting quietly					
grooming					
eating					
foraging (seeking food)					
chasing others					
vocalizing					

At first all the squirrels you watch may look pretty much alike. If you watch carefully at the same place for several days, you will begin to see things that help you recognize particular individuals. One may have a patch of hair missing from its tail; another may have a tear in its ear. When you are able to identify individuals, you will find it interesting to focus your observations on them. Follow one individual during each observation period. Compare your observations of different individuals. What activities and ways of doing them do they share in common? What activities are peculiar to certain individuals?

You may wonder if and how squirrels communicate with one another. Keep a record not only of the sounds you hear them make but also what they do with their bodies. Squirrels, particularly gray squirrels, communicate with their tails as well as their voices. Note down, or sketch, what they are doing with their bodies as they

Figure 4–1. *If you were a gray squirrel, how would you spend your day?*

interact with one another. If possible, note not only what one squirrel is doing but what other squirrels in the area are doing in response.

In much the same fashion you used with squirrels you can study birds. They are harder to follow because they can fly out of your viewing range easily and quickly. However, birds that set up a nesting territory can often be followed and observed quite easily if you are patient.

PROJECT: BECOMING A WILDLIFE DETECTIVE

Many animals you might like to find out about are either very shy or not active at the time when you can be. You can still investigate many things about their lives. Most creatures leave clues that they have been around. You can learn to find and interpret those clues.

Tracks are the clues most people think of but there are many others. These include such things as:

- bits of fur or hair stuck on rough surfaces
- uneaten remains of food
- undigested food residue that has passed through the animal and is known as scat or droppings
- flakes of bark chipped off where an animal has climbed
- tooth marks on twigs and branches
- nests and burrows
- pellets regurgitated by owls and some other birds

An interesting project is to make collections of items that will help you interpret some of the clues. For example, collect seeds from the plants that grow abundantly in your area. Spread them out to dry thoroughly and then store them in little plastic envelopes or clear plastic vials. Put the name of the plant they came from inside the containers. When you find a mouse nest with seeds in it, a squirrel's dining spot, or animal scat with lots of seed embedded in it, compare those seeds or nuts with your reference collection. When you find a match you will have part of an answer to the question, What does this animal eat?

Similarly you can make collections of hair and fur from animals that have been killed by cars or hunters or even your cat. These labeled samples you can then use to compare a mystery hair you have found caught on tree bark or on rocks or roots at a burrow entrance. Birds molt their feathers every year. A feather reference collection is useful to help identify other feathers you find in your wanderings. You can get started with game bird feathers supplied by hunters. To keep the feathers in good condition, glue them by the thick shaft at the base to Popsicle sticks. Write the name of the bird the feather came from on the stick. The stick will give you an easy way to handle and observe the feather without damaging it (Figure 4-2). Your reference collections of fur and feathers will help you answer the question, What animal passed here? Other kinds of reference collections to help unravel mystery clues will be detailed in chapter 8.

Insects and spiders are our most abundant animal neighbors. Usually they are the ones we know least well. We tend to ignore them unless they bother us by biting, stinging, or eating things we believe are for us alone. They are far more different from us than birds or mammals or even reptiles and amphibians. They seem too alien to get to know well. But when you take the time and effort to interview these small beings you will find that they lead interesting lives.

Sometimes all it takes is patient and careful observation. Two stories may help illustrate this. For a number of years my family raised dairy goats. The barn became a haven for flies. That was expected. However, it also became a haven for big white-faced hornets. My

Figure 4–2. *A set of labeled specimens is an invaluable aid to the "nature detective." Glue feathers you've collected to Popsicle sticks to create a reference collection.*

children were very concerned about doing their barn chores where these wasps were active because they pack quite a wallop with their sting. Instead of immediately turning to pesticides to stop the wasp menace I suggested we watch to see what the hornets were doing in the barn. What we discovered was that they were not there to molest us. They were there hunting flies! They would swoop down and catch a fly and carry it away to their paper nest in a nearby lilac bush. There it was paralyzed and stuffed into a nursery cell to provide food for developing young hornets. We learned that if we ignored them and minded our business they minded theirs. They cut down the fly population in the barn.

PROJECT: BUILDING CHOICE CHAMBERS TO HELP SMALL CREATURES ANSWER YOUR QUESTIONS

Asking questions of insects and spiders is at once difficult and easy. It is difficult because the creatures are small, often very active, and quick, and they flee or hide when observers approach them in the wild. They are easy because they can be kept captive in containers that let us create situations that get the animals to give us answers to some of our questions. To do this we need to create "choice chambers." Choice chambers are simply housing for the animals that is set up so that the animal chooses one part or another for most of its activity.

Perhaps the simplest such choice chamber can be set up to get caterpillars to tell you which plants they prefer for food. You will need a box tall enough to put in small branches of the plants you want to test (Figure 4-3). The cover of the box can be cut out and clear plastic taped over the opening so you can see inside the chamber. Cut out openings in opposite walls of the box and glue or tape pieces of old panty hose or cloth screening over them for ventilation.

You will want to set the box on jars or glasses of water. All the jars or glasses should be the same height so the box will set evenly. Over each of the containers you will punch holes in the box just large enough to push the stems of the food plants through. The stems will be pushed down into the jars of water so that the leaves will stay fresh.

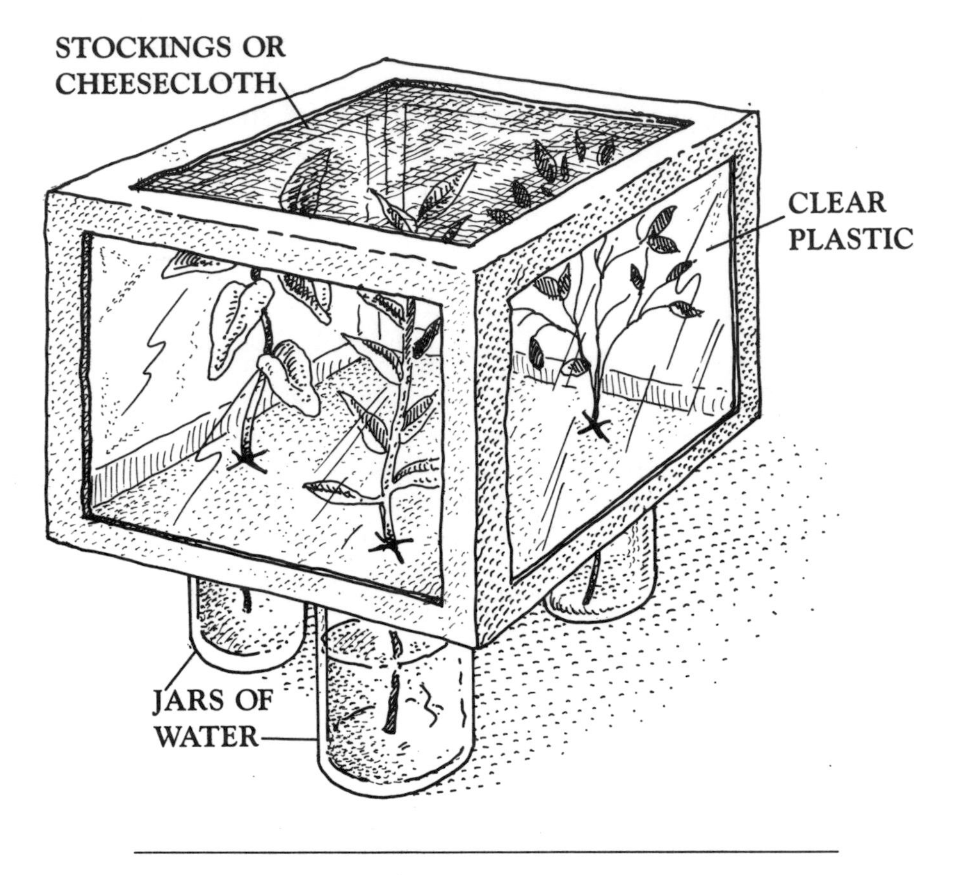

Figure 4–3. *A simple design for a choice chamber. Choice chambers help you to "interview" small creatures (caterpillars, for example) as to which plants they prefer to eat.*

Put the foods plants into the container, then add the caterpillars. Over the next few days observe which plants the caterpillars eat. They may climb up on a stem, nibble, and then move on to another. They will stay with those they really like and eat as many of the leaves as possible before trying something else. Some kinds of caterpillars will readily eat many different kinds of plants. Others are very fussy and will eat only a very few kinds. Often the different kinds eaten are all in one particular family of plants. The answers you get to what plants the caterpillars prefer can be very important to the survival of that species both in captivity and in your neighborhood in general. If those plants become scarce because of land clearing or other activity, the kinds of caterpillars that depend on those plants will disappear as well.

This type of choice chamber may also be used to answer another question you might ask of the caterpillars: Do you feed in the daylight or the darkness? If during the day you never see your caterpillars feeding but next day many leaves have been eaten, the caterpillars probably feed at night. You might ask them if it is darkness they prefer or do they have some kind of internal clock. If you cover the transparent window during the day with an opaque paper and peek in quickly from time to time do you see some of them feeding? If you do they are probably saying that it is darkness that they prefer for feeding. If you never see any feeding under these conditions, they probably are working on an internal biological clock. How would you interpret the finding that the caterpillars you are observing feed during the daylight hours even if you darken the container but do not feed during the nighttime hours?

There are other kinds of choice chambers that you can make. All are basically containers that can be divided into two parts. You then create different conditions in each half, add the creature you want to interview, and then record its behavior over a period of time. It answers your question by how it behaves.

Two designs for choice chambers are given here.

Using your imagination and materials you can locate, you can design and build similar working chambers.

Design one (Figure 4-4). Take two canning jar lids with removable disks. Discard the disks. Use two-part epoxy glue to glue the tops of the two lids together. Then screw on jars to each lid. Many other jars besides canning jars will fit the lids. This is the basic design of one kind of choice chamber.

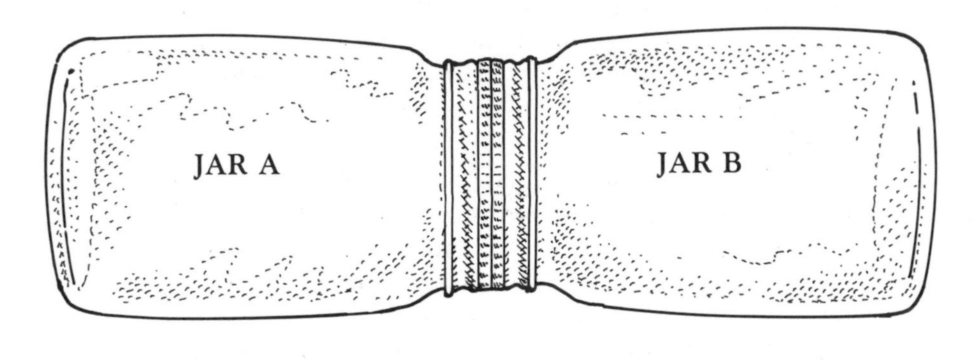

Figure 4–4. *Choice chambers made from canning jars. Using such chambers, you can investigate such questions as: Does the creature prefer damp ground or dry ground? Light or darkness? Warmth or coolness? Smooth or rough surfaces?*

Design two (Figure 4-5). Take a clear plastic sweater box with lid. Glue a strip of wood or plastic about one-half inch (1.25 cm) wide across the bottom of the box to divide it into two equal sections.

In either design add enough sand or soil to make the floor of the chambers level with each other. The animals to be interviewed can then easily move between the two chambers.

You can now state your question and then design the interior of the two sections of the chamber so that the creature can reveal the answer to you. You might ask: Do you prefer damp ground or dry ground? You would then moisten the soil in one half of the choice

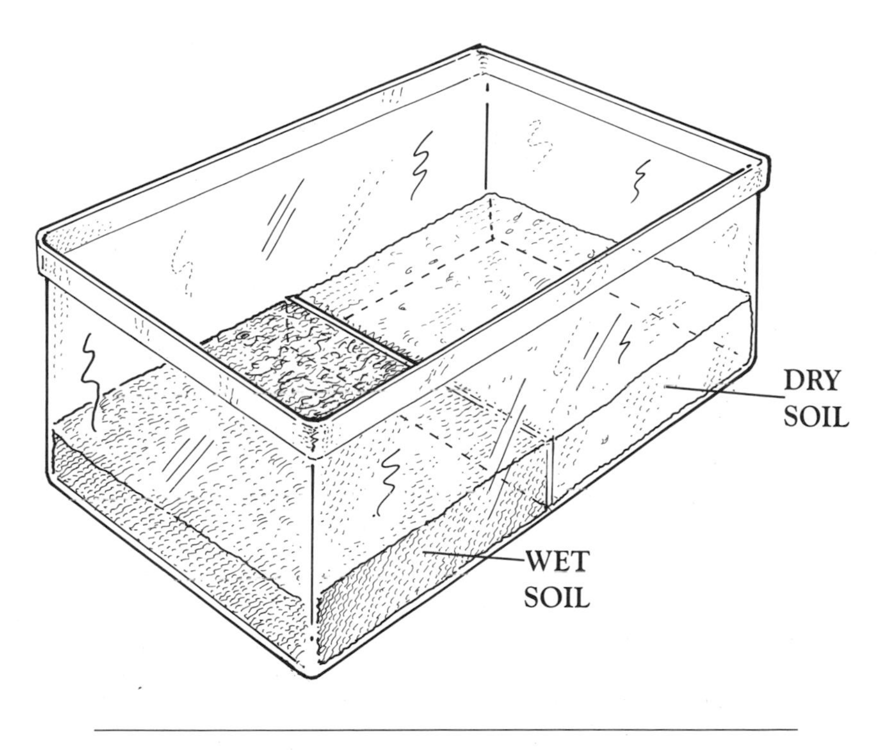

Figure 4–5. *Another type of choice chamber you can make*

chamber and leave the other dry. Place the animal you are interviewing in the chamber and give it time to become familiar with the whole chamber. After that time has passed, begin making regular observations and recording in which section you find the animal in each observation. Does it spend more time in one half of the chamber than the other? If so you might conclude that that is the condition it prefers. It may be found equally often in either half. In that case it may be concluded either that it has no preference for the moisture condition of the ground or that other things are more important to it that keep it moving around in the chamber.

Do you prefer light or darkness? Cover one half of the chamber with black construction paper, aluminum foil, or other opaque material. Then make your observations to get an answer.

Do you like it warmer or cooler? Set one half of the chamber on a heating pad, the other on a tray of ice.

Do you prefer smooth or rough surfaces? Put a piece of glass or plastic on the ground of one half and sandpaper on the other. Are rougher or finer sandpapers preferred? By now you should be able to figure out how to set that up.

If the question of moist or dry ground was not answered clearly and you suspect other factors might be involved you might try combining the light/dark setup with the wet/dry setup. You will want to do this in two ways. First put the dark over the wet side; repeat the observations with the dark over the dry side. Look at the information you collect. Has your interviewee given any clearer answers than when the two questions were asked separately?

The answers you get to these questions, or any others that you can devise only tell you what that individual prefers. It does not necessarily tell you what most members of its species prefer. You will have to ask the same questions, in the same way, of a number of individuals of that species to see how consistent the answers are to the questions if you are to make accurate statements about the species.

You can use choice chambers with a wide variety of small animals. It works well with insects and spiders but also with snails and slugs or salamanders and frogs. They even work with small mammals like mice and hamsters. One naturalist asked the question of two kinds of native mice, What thickness of plant stems made the best habitat for them? He used something very much like the sweater-box design. He drilled many holes in the lid and pushed different numbers of sticks through the holes into the sand on each side and let the animals show which density of stems they preferred. They told him that although they both would use the same areas at certain densities, one kind preferred greater densities and the other much lesser densities than in the area they both would use. This helped him understand why certain places had one species and not the other while other areas were home to both kinds.

PROJECT: CALLING IN WILDLIFE

In order to have animals to observe and ask questions of you have to either go to where they live or bring them close to where you can observe them. This generally involves one of three things:

- hiding yourself and calling the animals to you
- providing feeding and/or nesting areas that attract the animals nearer to where you can easily watch them
- capturing them and providing them with as many of their basic habitat needs as possible in small enclosed spaces. This is really only appropriate for small animals and requires a great deal of sensitivity and responsibility on the part of the naturalist. This is true even for such enclosures as choice chambers.

There are many ways to conceal yourself from animals you want to observe. Some are very simple; others quite complex. Many creatures will ignore motionless objects that they regularly come across in their habitat. Thus, having a regular observation place where you can sit motionless for a long period of time may be all you need. In fact, being motionless is not always necessary if you are persistent and patient. If you sit in the open and go about activities that the animals learn are not threatening to them such as reading or writing, some animals will ignore you and go about their lives.

Most creatures, however, are not so trusting or bold. You can make a simple blind, or hide as the British call them, in a variety of ways. As a youngster on a farm I used to pile hay bales to make a hide from which I could observe the local fox den and watch the young foxes at play. Branches can be cut and stuck into the ground to make a screen with places clear enough for observing. Some people just sit or lie in an observation point and cover themselves with a large piece of camouflage cloth that has some eye slits cut so they can peer out. Others stick stakes in the ground in the form of a square and wrap camouflage cloth around the four stakes to create a little room from which they can look out through slits cut in the cloth. Some cloth over the top helps prevent birds passing over from seeing in and sounding an alarm.

Fancier, more permanent blinds may be made from the crates used for shipping large appliances, like refrigerators or washing machines. Pop-up tents are adaptable as blinds and some very fancy, portable, tentlike blinds are made for use by wildlife photographers. Both of these are expensive, however.

There are various ways you can call wildlife closer to where you are hidden. Perhaps the simplest of these is making a squeaking sound by kissing the back of your hand. This may attract some birds and predators like foxes and coyotes. Many songbirds will also investigate a simple *pishing* sound with the *sh* sound prolonged and repeated several times in a row.

If you have a portable cassette tape recorder or player you can play the songs of birds available on cassette tapes. You will want to use the song of birds you hear around you. Males on their territory will come close to chase the rival that they think is inside your recorder. Don't overdo calling up any particular bird, as this may disturb its ability to hold a territory and raise a family. Distress calls of rabbits and young predators are also available and can be used to call foxes, bobcats, weasels, and other predators that may be in your area.

There are also available a wide range of mechanical calls (Figure 4-6). The Audubon bird caller is a rosin-

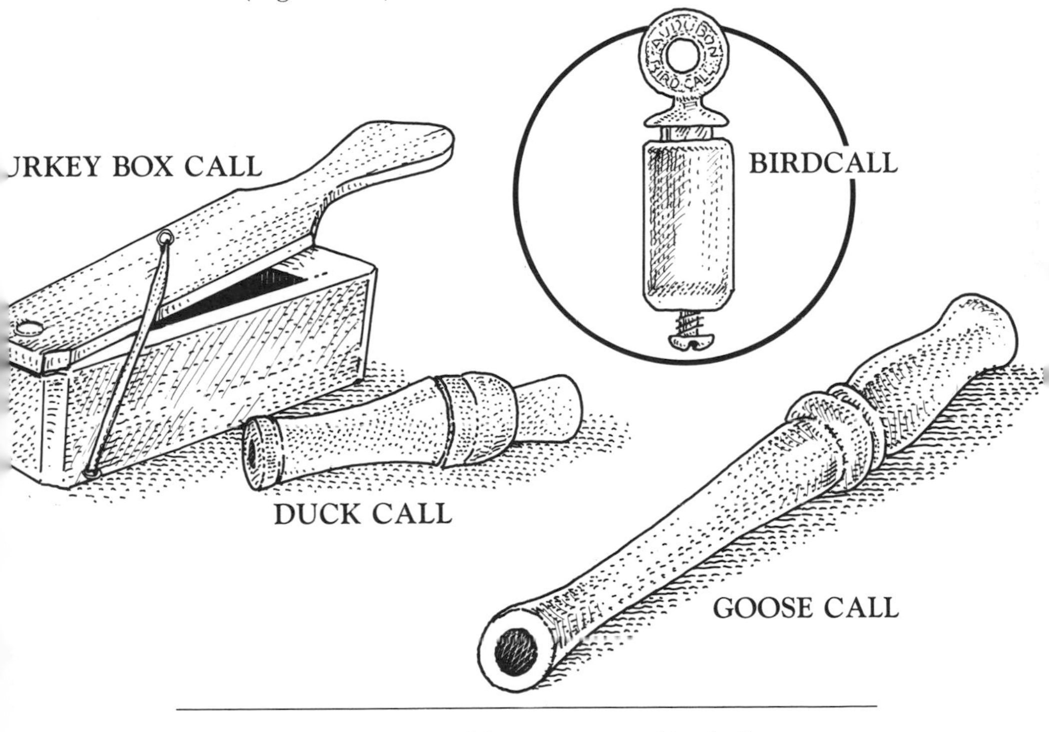

Figure 4–6. *Various types of birdcalls*

coated wooden cylinder that turns inside another wooden cylinder to make a squeaking sound. Turkey calls are a variation of this principle. Duck and goose calls are reeds like those found in some musical instruments, and there are also predator calls built on the same principle.

When calling animals to you it is important to be as quiet and motionless as possible, for the approaching wildlife will be on high alert for potential danger. You also must position yourself so that the likely approach of any animals is from the same direction any breeze is blowing. This will reduce the likelihood that your scent will be carried to them and frighten them away.

Asking Questions of Plants

Most people are easily attracted to animal life but they look at plants differently. It's harder to think of them as living things. They don't move about much and they don't seem to communicate with each other. Actually plants are a bit like animals in slow motion. Because the pace of plant lives is generally slower than those of animals, and because they tend to be stuck in one place all their lives, many people think of them more as features of the landscape than as living beings with fascinating life-styles.

If you take the time to examine their lives closely you will find that plants can be as fascinating as animals. Both face common problems of finding enough of the resources they need for living; avoiding attack by enemies, and of reproducing their kind. As an amateur naturalist you can help reveal the details of these life-styles both to yourself and others. Plants do have the advantage that generally they do not run away or do things so fast that if you take a time-out you might miss key action.

People find it easier to get to know others on a one-on-one basis. When you recognize a particular individual you can more easily focus on what that individual is doing. Likewise you will find it more interesting to follow the lives of individual plants and record what is happening in their lives. Fortunately, once you have located an individual it is not likely to go very far very rapidly. However, you may well find that some ferns and other individual plants move some distance over a period of several years.

PROJECT: FOLLOW THE ANNUAL LIFE HISTORY OF YOUR FAVORITE PLANT SPECIES

Like us, each plant lives a life that is unique to it. To know all about one individual is interesting but it does not allow you to suggest that all plants of the same kind live the same kind of life. To be able to do that you will need to follow the lives of many individuals of the same species of plant. You can then figure out what characteristics are true of all the individuals and which apply only to one or a few of them.

Choose a place where the plant species you are interested in grows and that you can visit and revisit frequently. Mark off an area that includes a number of the kind of plants that interest you. The size of the area will be determined by the size of the plant you are investigating. For small plants like dandelions an area about one yard or one meter square is good. For larger plants like trees you may choose an area several yards or meters square. Mark the boundaries of the area with stakes, rocks, or other markers that you can recognize easily when you return. It is a good idea to keep the markers fairly inconspicuous to others so that your plot is less easily vandalized.

Make a map of this area and locate each of the individual plants you will investigate on the map. Give each individual a number, letter, or even a name if you are so inclined. You may find it useful to put the letter, number, or name on the end of a Popsicle stick in indelible ink and stick this into the ground near the appropriate plant. Some people use the plastic markers favored by greenhouse workers. Some use little cardboard tags on strings available at stationery stores. On your map indicate the key compass points, for you will want to know the path of the sun across the area.

Create a life-history journal for your plants. In this journal you will have a space for each of the individual plants you are investigating. For each visit you make to the site you will want to record information about what has happened to each individual since you last visited.

Your first journal entries will focus on a description of each of the plants. You may start with big plants but you should also look for seedlings or even just seeds that you find on the ground. Record how big the plant is, how many leaves it has, what direction, if any, it is leaning in, what its plant neighbors are, and any other things you think may be important or which relate to questions you have about the plants. It is useful to make careful measurements so you can tell how much it has grown since your last visit; how much may have been eaten by enemies of the plant; and the like.

You will also want to record any insects you see on each plant or other creatures that you observe visiting it such as birds, squirrels, and the like. The more information you record about each plant the more you will eventually know about the lives of these beings. It is useful to keep a record of the weather in the area between visits. Plants are very responsive to the amount of rainfall, sunlight, wind, and other features of the weather. You may be able to correlate such things as amount of growth since the last visit with these weather features.

Follow the lives of your plants for at least one growing season and keep up the journal for each individual. During the winter season you can compare the lives of each individual and determine what activity was common to all and which were limited to only a few individuals. What you find out will probably trigger a lot of questions in your mind. You can begin exploring some of these questions with the same kinds of plants next season. You can design some simple comparison studies that will help the plants to answer your questions.

PROJECT: CREATE A BLOOM CALENDAR FOR LOCAL FLOWERS

For many people the first things that interest them about plants are the flowers. It is fun to keep a calendar of flowering for the many different plants that interest you. You can walk the woods and fields looking for the very first individual plants of a species to produce a blossom. This is like keeping track of the first robin of spring, or the first calls of spring peepers. Keeping track of such events is called phenology.

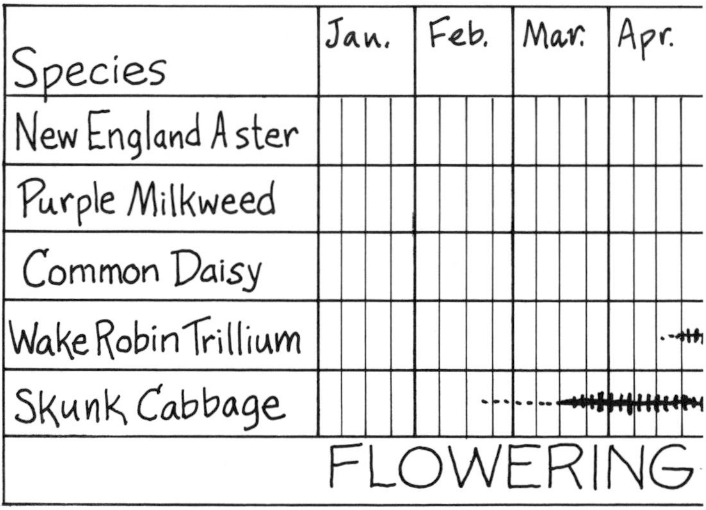

Species	Jan.	Feb.	Mar.	Apr.
New England Aster				
Purple Milkweed				
Common Daisy				
Wake Robin Trillium				
Skunk Cabbage				
	FLOWERING			

Figure 5–1. *A plant bloom record*

Good phenological records include more than just noting the first and last blooming individuals. In many ways such individuals are the oddballs. It is important to record as well when the most individuals of the species are in bloom. To get good records it is useful to set up study plots that include a large number of individuals of a particular species. Within the study plot keep track of when the first blooms occur and the last. But also record when a quarter, then a half, and then almost all of the plants in the study plot have bloomed.

The range of blooming times is important to the overall survival of the species. If there are later frosts, for instance, the first bloomers may not get pollinated and set seed. If there are early frosts or severe summer droughts, the later bloomers may not produce offspring. On the other hand, if there is a bad normal blooming

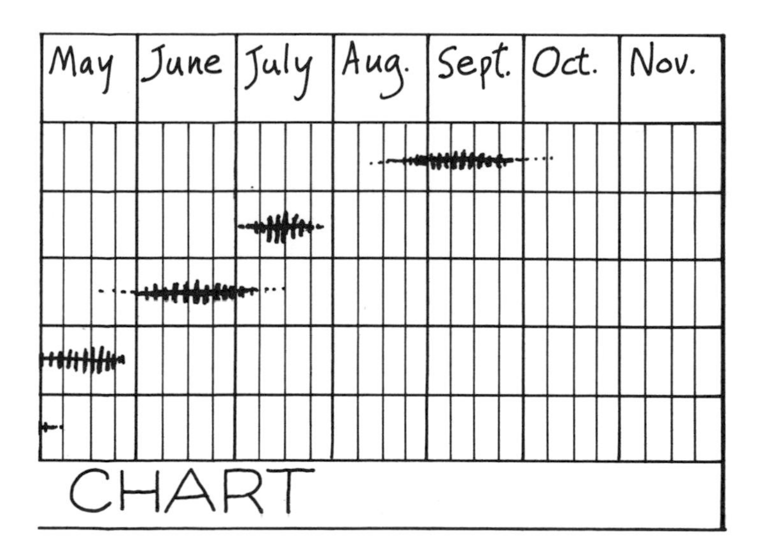

season only those who bloom early or late may successfully produce seeds that summer. A broad blooming season is one survival strategy that a number of plants use.

You can make a plant bloom calendar by keeping your records and then putting the information on a calendar using lines of expanding width like those shown on pages 56 and 57 in Figure 5-1.

For many plants, flowers are an important way to reproduce themselves. It is from the flowers that the seeds are formed. Many plants can reproduce without flowers and seeds. They grow from bulbs, runners, rhizomes, and other underground parts that send up new growth. Sometimes these ways of reproducing are more common for a plant than reproduction from seeds. A whole mountainside of aspen trees may actually be the underground reproduction of one plant. The same may be true of a large clump of sumac bushes or blackberries.

PROJECT: KEEPING TRACK OF PLANT VISITORS

However, flowers are very important for the reproduction of most plants. In most cases it is essential that the pollen be removed from one flower to another for there to be the fertilization that will result in a viable seed. Over many generations plants and a variety of small animals have co-evolved relationships that ensure that the pollen gets moved. We commonly think of bees as the primary creatures that do the job. It is true that they are important but so are many other creatures, including moths, flies, hummingbirds, and in many parts of the world, bats.

It is fun to spy on plants and find out just what insects are visiting them regularly and what they are doing on the plant. Keep a record of what insects and other creatures regularly visit the plant. Are they frequent or infrequent visitors? Do they visit the plant only while it is in flower? Do they behave in particular ways when they visit, such as always landing on a particular part of the plant? Do they come for the pollen or something else? How is the plant designed to lead the visitor to getting the pollen on them? How do they get pollen on themselves and transport it to another flower? Is it always the same kind of flower that they last visited?

What other reasons do the visitors have for visiting the plant besides the plant's need to be pollinated? Do they use the plant as just a convenient landing platform? Do they seek food? If so, what part of the plant are they feeding on? Or are they seeking other visitors to the plant to eat? Do they come to lay their own eggs and reproduce their kind? These are the questions for which you can gather answers as you observe the comings and goings of animal visitors to the plants and their flowers.

PROJECT: GROW NEW PLANTS FROM SPORES

Of course not all plants produce seeds. Many plants produce spores instead. These plants include ferns and mosses. Another group of living things, the fungi, which include the familiar mushrooms or toadstools, also produce spores. Fungi were once considered to be a kind of plant, but today they are placed in a separate kingdom of their own. One reason is that none of the fungi have chlorophyll, the green pigment that helps true plants trap the energy of the sun in the form of simple sugars. It is a challenge to grow spores into adults, but it can also be fun and there are some real surprises, for most spore-producing plants live two quite different-appearing lives.

A key ingredient for growing plants from spores is patience. On average, it takes six to ten months from the time you sow the spores until the young plants can be safely transplanted. Ferns are probably the best spore-producing plants to try and grow.

If you can't find ferns growing wild locally, ask a local florist if he or she can find you a fern frond with spores among the fern plants in stock. Spores are found on the underside of the leaflets of many species of ferns in little spore sacs called sporangia. A few kinds of ferns put up special stalks that are completely covered with sporangia but have no spore sacs on other leaves.

Collect the fronds with sporangia when the sporangia are light brown and have the skin covering them intact. If the spore sacs are shriveled or frayed the spores have already been shed. If the skin covering the sporangia is green, the spores are not yet ripe.

Put the fronds you collect, sporangia side down, on a sheet of paper. Cover the fronds with a jar to keep the spores from blowing away. Leave them there for several days. When the spores are ripe, the skin of the spore sac will rupture and the spores will fall onto the paper.

To grow the spores, the inverted-pot method is a good one (Figure 5-2). Take a clean, porous clay pot that is not too large to be completely covered by a glass jar. Fill the pot with sphagnum, or peat moss. Then invert it into a shallow dish or saucer. Pour boiling water over the pot and saucer several times to sterilize them. Drain the water away and cover the pot with the glass jar and let everything cool while you wait for the spores to ripen.

When you can see some of the brown spores on the paper underneath the fern frond you are ready to plant. Carefully lift the glass off the fronds. Avoid creating a breeze that will blow away the dustlike spores. Gently lift the fern frond and remove them, leaving the spores on the paper. Put the jar back over the spores and carefully move paper, spores, and jar to the sterilized pot.

Gently remove both glass jars. Hold the spore-cov-

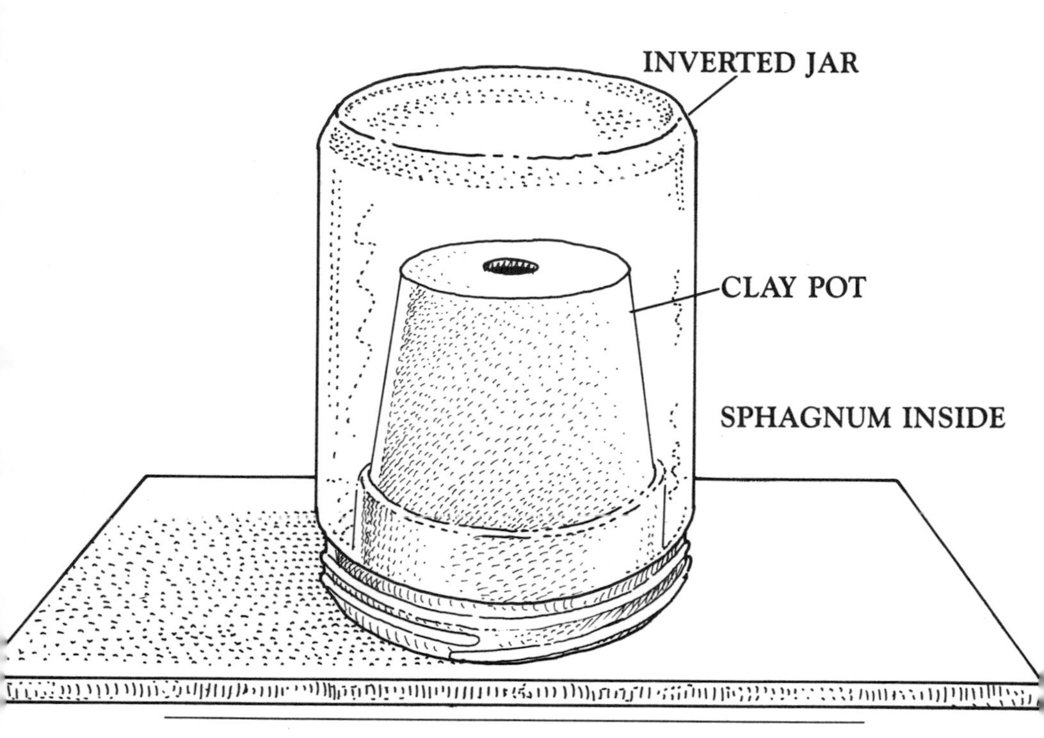

Figure 5–2. *The "inverted-pot method" of growing plants from spores*

ered paper over the flowerpot and tap the spores off the paper onto the pot. Then cover the pot with its jar. Place the pot and jar in a place with a low to medium intensity of sunlight, or put the setup under a floures-cent lamp where it can receive eight to sixteen hours of light per day. Keep temperatures between 68 and 86 degrees Fahrenheit (20 to 30 degrees C). Keep the pot moist. Watering with a mister is better than a watering can. It is less likely to wash the young plants away. It is best to use distilled water or boiled water that has been cooled. A weak solution of liquid fertilizer can be added to the water every couple of weeks after a green mat appears on the pot.

The green mat can be examined with a magnifying glass. The little plants that you see will not look anything like ferns. They will be thin, ribbonlike structures. Each is called a prothallium. It is the non-sexual stage in the life history of the fern. (It does not grow from the joining of male and female plant cells.) On the underside of the prothallium are some threadlike structures that hold it in place. They are like roots but do not have the same structure.

What is truly remarkable about the underside of the prothallium is that it will develop two strange-looking structures. One looks like an upside-down flask; the other like a little pimple. The flask-shaped one will develop a female cell or egg. The pimple-shaped one will develop spiral-shaped male cells. If moisture is present the male cells will swim to the female cells and they will join together to form what is called a zygote.

It is the zygote whose cells will divide over and over again and grow into the fronds that we recognize as fern plants. The fronds will then eventually create the spores that will grow into prothallia. Spores do not form from separate male and female cells. They are considered to be asexual. What makes ferns and mosses and their relatives so unique is that they keep alternating generations. The spores grow into prothallia asexually but they produce two sexes that join to form the second generation sexually. This second generation in turn produces more spores asexually. And so it goes, back and forth, generation after generation.

Once a mat of prothallia has formed on your clay pot you can transplant them to larger quarters. Put some sphagnum peat in a clear plastic container with a lid. The sphagnum should be sterilized with boiling water and cooled. Also drench the sphagnum with liquid fungicide, which you can get at a local plant supply store. Cut quarter-inch (6-mm) pieces of prothallia from the clay pot and press them into the peat. Space the piece of prothallia about a half inch (1.25 cm) apart. Mist them with distilled water. Cover the containers and continue to keep watch.

As young fern fronds begin to grow, you may want to divide them again into larger containers. Once they are a few inches tall you can remove the cover and let the young fronds toughen for several days to a week. They can then be transplanted to individual peat pots. Later you can transfer them outdoors or to indoor or patio planter boxes.

Once you are successful with ferns, you may want to try and grow other spore-bearing plants, like mosses and club mosses, from spores. You may also try growing mushrooms. Some garden supply companies sell spores of edible mushroom varieties with directions on how to grow them.

PROJECT: FINDING OUT HOW MANY SEEDS SURVIVE TO ADULT PLANTHOOD

A question to ask the plants, particularly the trees in the neighborhood, is, How many of your seeds ever become young plants? Each year it seems the trees in the neighborhood grow many, many seeds. What happens to them? You can undertake a project that will help answer this question.

You will notice that tree seeds have a variety of ways to move about. Acorns from oaks, and other kinds of nuts, are heavy. They drop like bombs from the parent tree branches and land under or not far from the parent. Others, like maples and elms, have wings on their seeds that catch the wind and help carry them away from their parents. To explore the fate of seeds of such trees you need to find out what is the most common wind direction. You can then figure out the pattern in which most of the seeds are likely to land. There will be a few seeds upwind of the tree and many downwind.

To find out as much as you can about what happens to the seeds you will need to set up study plots. You can make them any reasonable size and shape you want. Circles are good for acorn- and nut-producing trees. Rectangles may be more useful with wind-dispersed seeds. Mark the boundaries of your plots with Popsicle sticks or plant stakes pushed securely into the ground. When you are working your plot you can run string around these stakes to more clearly show you the study area.

Make a scale map of your plots. It is helpful to use graph paper to do this. Locate each seed in the plot on your map. Give each seed a number. In your naturalist's journal keep a record of what happens to each seed. Your first entry in the journal for each seed should tell you something about the conditions where the seed is located. Is it on a rock? On sand? On moist soil? In a crack? In the shade of another plant, etc.?

Visit your plots on a regular basis and record what is happening to each of the seeds. Which ones sprout and when? Which ones are missing? Can you tell why? Did they blow or wash away? Did some animal collect and eat them?

At the end of the growing season, check over your journal. Of the number of seeds that were in the plot when you started, how many actually sprouted? Of those that sprouted, how many survived and are still alive and growing? You can check them again at the end of the next growing season to see how many survived a second year. You may then also want to see how many new seeds fall into your study plots and how many of them survive their first year. As you can see, this is an exploration that can go on for many years and will help you find out just how difficult it is to be a plant.

We have focused here on trees but if you are an alert and careful observer, you can follow the seeding and survival of almost any plant species that interests you this way. You have to be able to recognize that plant's seeds, and that may require some very careful observation ahead of time.

PROJECT: MAKING YOUR OWN REFERENCE COLLECTION OF SEEDS, LEAVES, AND WHOLE PLANTS

When you examine a plot for the seeds that are there you may find many more kinds of seeds than you expect. It is not likely that you will be able to find a field guide to help you identify them. Your best identification aid would be a reference collection of seeds from plants that you have identified earlier. It is much easier to find a field guide to local trees and wildflowers.

Once you have identified a plant, keep track of it until it produces seeds. Collect some of the seeds and put them in small clear glass or plastic vials or the glassine envelopes often used by stamp collectors. The tops should be left off for several days or a week to let the seeds dry well. A small amount of a drying agent such as silica gel or a calcium salt should be put in the containers before you seal them in order to absorb any moisture that is in the air or still in the seeds. You can get such material at florist or craft shops. The material is used by craftspeople to dry flowers.

Label each seed type well. One label with the common and scientific names should go inside the container but it may be hard to read there. Therefore you should also put a similar label on the outside. The reason for two labels is that the outside one, while easier to read, may come off. With the inside label as backup, you are not left with an unlabeled specimen.

If you find a seed out in the fields or forest that you do not recognize, collect a sample and compare it with your collection of specimens until you find a match. You will discover that in the process of making the seed collection you have taught yourself to recognize a great many seeds without having to first compare them to the actual labeled specimens.

Making a labeled leaf collection can also be an aid to plant identification. You can also have a great deal of fun by creating art projects for leaf collection or just plain decoration. Collect the leaves and identify them with a field guide. As leaves dry they get very brittle and crumble easily. To preserve them, place the leaf on a sheet of white paper. Cut a rectangle of clear contact paper. Peel off the protective backing and put the sticky side over the leaf and rub it well onto the paper and the leaf. Label your specimen carefully and tell where you found it. You can store your specimens in a three-ring binder. If you get unwanted air bubbles, prick them with a needle to let the air out and press down the contact paper.

Some people preserve just the image of the leaf. There are a number of ways to do this.

- Hold a pane of glass above a burning candle and get it covered with soot (Figure 5-3). Lay the leaf on the soot and rub gently, getting soot on the veins and the edges of the leaf. Place the leaf, sooty side down, on a clean sheet of paper. Rub gently. Remove the leaf very carefully. A print of the leaf should be left behind on the paper. It will smudge easily. Spray it with clear plastic spray or the fixative used by artists to "fix" pastel crayon drawings.
- Similarly, you can rub the leaves on an ink pad and then put the ink-covered leaves on paper and rub. The dried ink does not require a fixative. The problem is that most ink pads are too small for many leaves. An alternative to an ink pad is to use an artist's roller to spread paste ink on a sheet of glass. Treat this as if it were

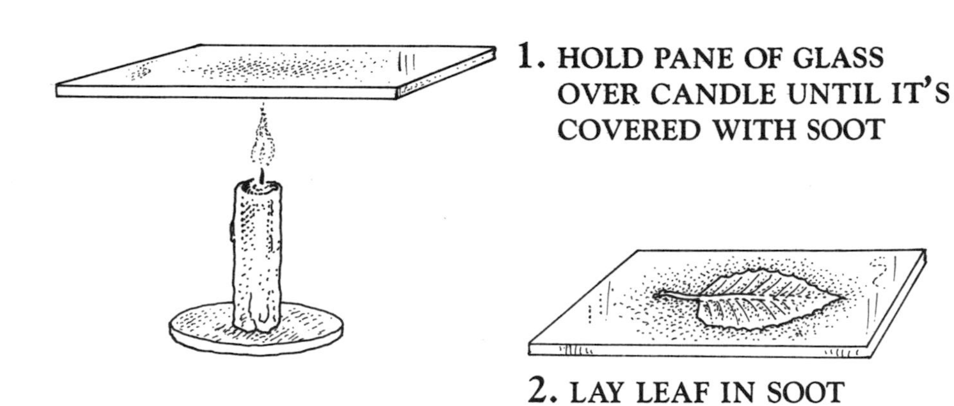

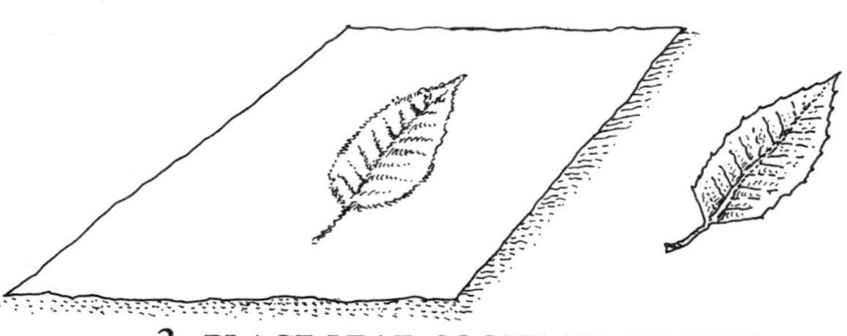

Figure 5–3. *One method of preserving the image of a leaf. (Spray the image with clear plastic spray to prevent smudging.)*

an ink pad and coat your leaf and print it on paper.

• Another way is to put the leaf on paper and then gently spray paint over it. This can be done with colored, or black, ink in a pump sprayer; a can of spray paint; or by dipping a used toothbrush in ink or watercolor paints and flicking the plant with your thumbnail or rubbing the brush over a piece of screening. Both ways spatter the paint over the leaf. When the ink or paint is dry, remove the leaf and you will have an outline of its shape.

Using these techniques, you can make designs on the paper with your leaves, creating note paper and other items as well as your reference collection.

To make the most useful reference collection of plants you will want to preserve the whole plant if possible. That is difficult if the plant has a thick, tuberous root. In such cases you will preserve the rest of the plant and make a careful drawing of the root on the finished specimen.

You will need to collect only one or two plants of each kind for your reference collection. Take a large, self-sealing plastic bag with you when you go collecting. Put your specimens in it and reseal it to keep them from drying out while you are afield. It is useful to get into the habit of tying a label on the specimen, giving it a collection number and noting where it was collected.

Back home the specimens are dried and pressed. This is done by putting the plants between layers of newspaper, paper towels, or desk blotter. These will absorb the moisture. Then you will put the sandwiches of plant and paper under pressure.

When you put the plant into the sandwich think about how it will fit later on a sheet of plain 8½-by-11-inch paper. You will have to bend taller plants into zigzag shapes in order for them to fit. Do this before you dry the plant! It will be brittle after it dries and break if you try to bend it too much.

You can get special plant presses to squeeze the moisture out of the plant but they are not necessary. Old phone books, the thicker the better, can be recycled for pressing plants. Stick the plant/absorber sandwiches between the pages of the phone book and then put something heavy on top like a stack of heavy books, bricks, or concrete blocks. Every few days change the absorbent materials.

When the plants are thoroughly dry mount them on clean white paper and label them. You can tape them down to the paper with transparent tape. It will come loose after a few years, however. You can also spread white glue on a sheet of glass. Then lay the plant carefully on the glue. Pick it up carefully and lay it on the paper and let it dry. The white glue will dry clear. The glue will wash off the glass with water.

The label on your specimen should include:

- Location where the specimen was collected
- Date it was collected
- The name of the person that collected it
- The name of the plant: both its common and its scientific name
- Name of the person who made the identification

PROJECT: PLANT MYSTERIES QUEST

The plant world is full of marvelous mysteries. Some of them we know a fair amount about, others much less. Many of these phenomena are reasonably common and the curious naturalist can set out to discover them and explore them personally.

Fairy Rings • Often when there is a fairly large open space with mushrooms growing you may notice that they seem to be growing in a large ring (Figure 5-4). There are no mushrooms in the center; just clusters of them around the edge of a circle.

It is fairly common to have a broken circle. If you check the incomplete section you will usually find a rock, log, or other obstruction on or just below the surface. If you return to the same place year after year, you are

Figure 5–4. *Mushrooms often grow in large circles known as "fairy rings."*

likely to find the ring again but with an ever-greater circumference. You can check this by putting colored golf tees in the ground next to each mushroom so you can check the area again next year. Or you can find the center of the ring and put a marker there. Then measure from that to each mushroom and make a map. Next year when you find the ring measure, you will map again and plot this map over the first one and so on year after year.

The secret of fairy rings lies in the growth habits of many fungi. Mushrooms, known to botanists as carpophores, are the fruiting body of the fungi. They develop to create the spores that give rise to a new generation of that species. The main body of the fungus is a mat of white threadlike bodies called hyphae. The whole mat is known as the mycelium. The hyphae grow only at their tips and the mat tends to become circular in shape. At the proper time hyphae at the outer edge fuse together and develop into the mushrooms. These appear above ground. Since the mycelium is essentially round in shape, the mushrooms appear to grow in a ring or circle.

When the spores have been shed the mushrooms die back. Underground the hyphae keep growing outward from the circle. Next year they will again fuse and form a ring of mushrooms. This ring will be farther away from the center because of the amount of growth the mycelium has had during the year. If the mycelium reaches a place where it cannot grow farther or is actually damaged, that section will not produce new mushrooms. The ring will be broken. In time most rings break up into wavy or curved lines.

Foxfire • A summertime walk in a moist woods in the dark may give you a surprise when you come upon a log glowing in the dark. This is what is known as foxfire. You can take chunks of the moist wood and bring them home and they will continue to glow in the dark for many days if you don't let them dry out. Conditions have to be just right for foxfire. You will have to be patient in your search.

If you examine the wood with a powerful hand lens you will discover the threadlike hyphae of a fungus. The fungus is "digesting" the dead wood, helping it decay. In the process the fungi creates a waste product that is luminescent, that is, it glows in the dark. There is a fungus, called the jack-o'-lantern fungus, that will glow in the same way on the underside of the mushroom. It is likely that the foxfire is the mycelium of that type of fungus or a close relative.

Jack-o'-lantern mushrooms are usually found at the base of a tree. By the time they appear the mycelium has probably killed the tree. This is why it is likely that this fungus is the major cause of the phenomenon we call foxfire.

Exploding Seed Pods • Much more common than foxfire are seed pods that "explode" and send their seeds some distance from the parent plant. A fairly common one to look for is jewelweed. Look in moist rich soils. The stem of jewelweed is fairly thick and translucent (Figure 5-5). The leaves are oval with wavy edges. The flowers are orange or yellow, thimble shaped with a little curled tail.

These flowers, which are favored by hummingbirds, develop into the seed pods after fertilization. Each pod normally has two seeds. As the pod ripens and the seeds mature turning from green to brown, it swells larger and larger. It gets very taut. At the slightest touch it ruptures and an internal coiled "spring" flings out the seeds.

It is fun to touch these pods and attempt to catch the seeds. They can be eaten and taste like black walnuts. Of course you can also put a paper under the pods and touch them with a stick and let the seeds land on

Figure 5–5. *The jewelweed plant has orange or yellow flowers and is known for its exploding seed pods.*

the paper. How far can the "explosion" shoot the seeds away from the parent plant?

Along railway rights of way and other shrubby places look for a vine among the bushes that has an oval-shaped spiny fruit. This is the squirting cucumber. Bring it inside and let it dry and see how far it shoots its seeds. You can do the same with the fruits of witch hazel, a fall-flowering shrub.

Pollen Catapults • Some flowers have both male and female parts. The pollen from the male stamens has to be able to get to the female stigma in order for pollination to occur. Many plants count on insects, birds, or some kinds of bats to move the pollen around. Others use the wind. A few have developed mechanical devices to launch the pollen toward the stigma. You can observe this with mountain laurel and its relatives. In these flowers the stamens arch out from the center and the pollen-bearing anther is lodged in a little pocket in the petals of the flower. As the pollen ripens the pocket dries a little, releasing the anther, and the curved stamen acts like a catapult to pop the anther up and fling the pollen onto the sticky stigma. You can manipulate the pockets on the flower and watch the catapult action for yourself.

There are many fascinating seed-dispersal strategies to be found among the plants of your area. Make a collection of the different kinds and compare them to manmade objects that have similar functions. Can you find:

- parachutes or parasails?
- helicopterlike rotor blades?
- grappling hooks of various types?

Some plants live their whole lives in one growing season. Others live for two to three seasons. They put all their energy into making the leaves and structure of the plant in the first year and devote the second year to making flowers and ultimately seeds. Then they die. Still others go on living and growing year after year after year. The oldest living individuals on earth are trees. Some are several thousand years old. No matter what their life history pattern, plants tend to do things more slowly than animals and it requires patience and persistence on the part of the curious naturalist to get them to reveal the answers to questions about how they live their lives.

Asking Questions about Plant and Animal Habitats

It is easy to think of individual plants and animals. People seem to relate to them more easily that way. And we think of each individual as if it were an island completely apart from anything else. But an animal cannot exist without its food, which may be other animals and/or plants. It cannot live without water or shelter. Thus, these things are a real part of an individual. To know an individual well you must know its habitat as well. An animal in a cage is only a part of itself. A plant in a greenhouse likewise is only a piece of its whole self. The truly curious naturalist must explore a wide range of things in order to fully understand even a small part. To be a good naturalist you must be able to ask questions about a creature's habitat as well as of the creature itself.

People tend to view things only from the human vantage point. We view the world from our eye level. On a late afternoon we may be standing outdoors and feel a breeze on our face. The breeze evaporates the sweat on our face and we feel cooler. At the same time we are unlikely to notice that there is no breeze down at our ankle level. In fact the air down there is warmer than up at face level. The heat that was absorbed by the earth during the day is radiated out into the air.

If you are a grasshopper living in the grass this difference is very important. During the day you climb high up in the grass to catch a bit of breeze and avoid the extra heat of the earth. But you must not get too much direct sun yourself or you may overheat. At night when things cool you go back down the grass stems to the earth, where the radiation from the earth keeps the air warmer. You have no sweat glands like people do to help regulate your body temperature. You migrate up and down the grass to find comfortable conditions.

People think of climate in terms of average conditions at the levels in which we usually operate. Within such a broad range there are actually lots of micro-climates. These are areas and places where the average physical conditions are quite different. Such places include holes in the ground, the shaded side of rocks and trees, the air a few inches above the ground, places protected from the wind, and many others. These places often provide conditions that are comfortable for certain plants or animals that can't survive long even a short distance away.

For example, very few desert animals can survive the heat and dryness of desert daytime conditions. In fact many desert creatures, like the kangaroo rat (Figure 6-1), actually have a very narrow range of tolerance for heat and cold and moisture. Yet they survive in the desert. How? They use the desert's micro-climates. The desert cools dramatically once the sun goes down. The kangaroo rats choose to be active at night, while it is cool. The surface of the desert gets very hot but a few feet beneath the surface it is much cooler. The kangaroo rats dig their burrows down to this cool zone. Here they spend their days. Temperature and moisture in the burrow is much like that of the desert at night. So the temperature range they live in is actually very narrow.

These animals are also physically adapted so that they lose very little moisture from their bodies when they are active. They have very short, rounded ears, for example. They are also adapted so that their bodies can

Figure 6–1. *A kangaroo rat*

remove the moisture in their food by digestion and use most of it in their own bodies. They recycle the water over and over and release very little of it in body wastes.

How do we know these things? Naturalists have spent time with thermometers, humidity-testing devices, and other tools to record environmental conditions. They compare their findings with their observations of the animal's activity patterns. When the activity patterns match with certain environmental information they have a reasonable answer to the questions of how the animals can survive under the desert conditions.

Information on the adaptations comes from making guesses about the function of certain structures and checking these guesses against laboratory studies. This is how we have come to know how they reclaim and use water from their food and the relationships between the size and shape of their body parts to moisture loss.

Many desert-dwelling animals have developed similar adaptations and behavior patterns, although the details vary. Many plants have developed growth patterns that reduce their exposure to the most direct rays of the sun and adaptations that minimize water loss to the environment. A number have developed very long root systems that grow down to water stored deep underground.

You may not be studying desert creatures but you can check out the habitat conditions of the plants and animals you are investigating. In actuality the habitat is a very real part of every animal. No animal or plant is completely bounded by its outer covering of skin, bark, or membrane. It is interconnected intimately with its habitat. To know the creature truly, you must also know its habitat.

PROJECT: EXPLORING MICRO-HABITATS

Exploring micro-habitats involves taking measurements and recording data. You will need to find out what is happening at the particular site you have chosen to explore and compare that information with information from other points in the same general area.

Temperature Pole • For example, to determine what the temperature climate is like for animals that live on or near the ground, you will want to measure temperatures at different heights above the ground. You might choose temperature at ground level, one foot (30 cm), and three feet (90 cm) above. Temperature poles are useful tools for doing this (Figure 6-2). Take an old broom handle or similar pole. Sharpen one end so that it can easily be stuck in the ground. Fasten three empty cardboard tubes, such as those from empty toilet paper rolls or paper towel rolls, to the pole at the appropriate levels. To determine ground level, stick the pole in the ground so that it stands well without falling over. Mark the ground level on the pole. Remove the pole from the ground and attach the cardboard tube where you made the mark on the pole. Attach the other tubes at the appropriate levels above this.

You will then put your thermometers in the tubes. They will be well supported, but out of the wind and out of the direct sun. To begin with, put all the thermometers you will be using on a table and leave them for a half hour or so. Then check them all to see what they read. If one reads quite differently from the others don't use it. If you are using the type of thermometer that is held into the markings by staples, you may be able to adjust the tubes by sliding them about. Adjust them so that all are giving the same reading before using them in your investigations.

Put out your temperature poles in the places you want to investigate and place the thermometers in the cardboard tubes. Take readings from the thermometers at the same times each day. Your field notes might take the form of a chart with the following headings:

Location	Height	Time	Reading	Animal Observations

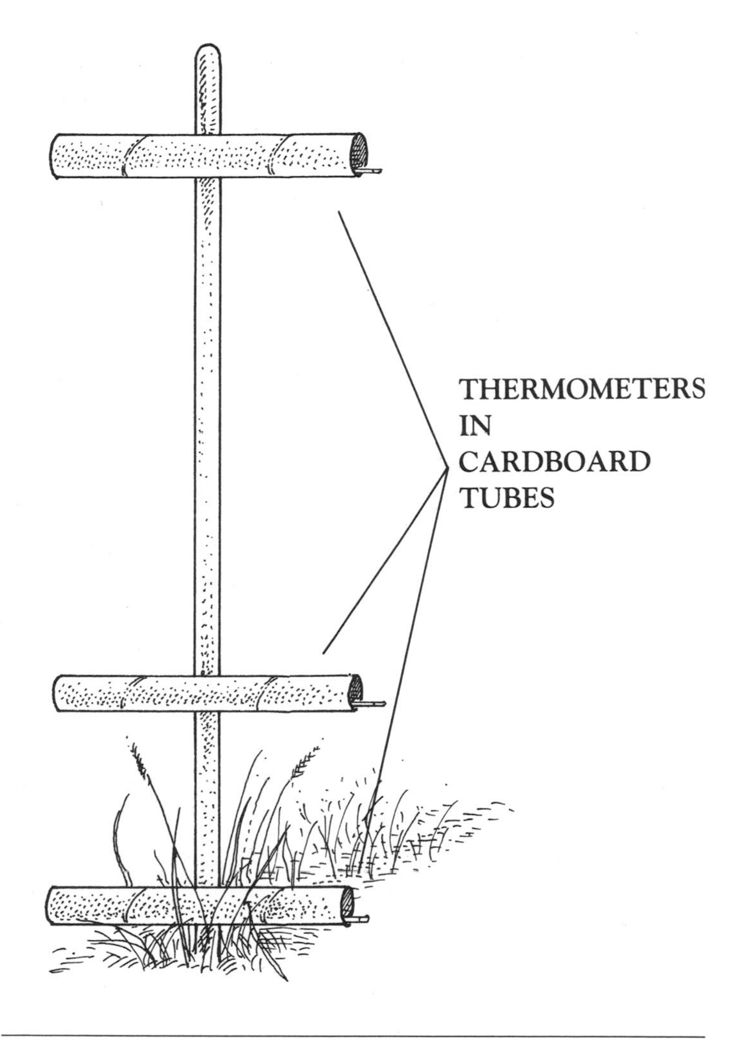

Figure 6–2. *A temperature pole is a useful aid in exploring microhabitats.*

Rain gauge • When you hear a weather report and it tells you that an inch of rain fell, that doesn't mean that every place got that much rain. Weather people take their measurements in the open. The ground under trees will receive much less water than the ground in open fields. This contributes to the conditions of a micro-climate. You can make and set out rain gauges to get a much better idea of how much moisture is reaching your study area.

To make rain gauges you will need several large plastic funnels. All should be the same diameter. This is so that they will each be collecting rainfall over the same-size area. The funnels should sit snugly on their water-collecting containers. These containers may be empty fruit juice cans or wide-mouth jars. All your water-collecting containers must be the same diameter. You may want to use several pieces of duct tape to be sure the funnels stay in place on their containers.

To be sure that your rain gauges don't blow over in a storm, you will want to make holders for them. These can be as simple as old broom handles that are pointed at one end so they can be driven into the ground or even plant stakes from the family garden. Large, wide rubber bands can hold the rain gauges to the stakes. If you want to take your measurements above the ground you will need to attach a platform to your tall stake so that the rain gauge can sit on it at the height that you desire.

To measure how much rain fell you will need to make a measuring jar. For this you will need a tall, slender, straight-sided jar with a flat bottom. These are the kind of jars olives often come in. The measuring jar should not be more than about 1½ inches (3.75 cm) wide or it will be difficult to get accurate measurements. Take one of your collecting jars, without its funnel. Carefully fill it with water to a depth of exactly 1 inch (2.5 cm).

Next pour the water into the measuring jar. Set the measuring jar on a flat, level surface and carefully mark the water line on the outside of the jar. To do this you can use a grease pencil, some nail polish, or model airplane dope. Put a good point on the grease pencil or use a very fine brush with the polish or dope to keep your line nice and narrow. Now use a ruler to divide the space between the line and the inside bottom of the jar into ten equal parts. Make marks for each $\frac{1}{10}$ of an inch (2.5 mm).

If you live in an area that regularly gets more than an inch of rain in any particular storm, you may use the same process starting with 2 or 3 inches (5 or 7.5 cm) of water in the collecting jar. You will still want to divide the spaces between each inch marker into ten equal parts.

Put your gauges out into the habitats you want to study. Be sure that at least one of the gauges is set out in an open area so that it can establish what the major rainfall for each particular storm is. You can then compare all the other measurements you get to this one. To measure, empty each collecting container into the measuring jar and read the water level to the nearest $\frac{1}{10}$ of an inch. To keep your accuracy high, dry the inside of the measuring jar with paper toweling between readings.

PROJECT: MAKING SIMPLE SOIL PROFILES

Soil is an important part of many habitats. Its texture determines how well it will hold water or let it drain away. Likewise, its texture determines how easily plant roots can push through it or how easily some animals can dig into it.

Soils are actually a mixture of rock particles, decaying plant and animal matter, air, and water. The kinds of rock in the mix and the amount of decaying material determine the kinds of nutrients that will be available to plants and many tiny animals. Soil is more than "just dirt." There are many different kinds just as there are many different kinds of plants and animals. Many plants and their associated animals prefer one type of soil over another. The distribution of some is strictly limited to specific soil types.

Soil types develop slowly over time. They begin with the weathering of rock. The rock breaks up through the forces of freezing, thawing, lichen acids, and related events. The particles are moved about through the forces of erosion—wind, water, and gravity. This means that the slope of the land will have a big part in shaping soil type. The decaying plant and animal material may come from things growing on the weathered rock or may be brought to the site by the same forces of erosion that moved the rock material. Over time the emerging soil develops on a site as distinct layers with more plain rock material on the bottom and more plant and animal material on the top.

Water working down from the surface dissolves surface materials and small particles and carries it downward. This may settle out a few inches or a foot or more beneath the surface, creating another layer. The different layers usually can be recognized by differences in color. Soil specialists call each of the different layers soil horizons.

The decaying material at the surface also shows fairly distinct layers. At the very surface is newly fallen or blown-in material that is easy to recognize. This is called **litter.** Beneath it is a layer of material that is made up of broken-up pieces of the litter. The pieces still resemble the nature of the pieces of litter they came from. This layer is called **duff**. Beneath the duff is a third layer of decayed material whose origins can no longer be recognized. This material is usually fairly dark and moist. It is called **humus** and is the source of much of the nutrient materials plants take from the soil.

The ability of the soil to hold water and air depends upon the spaces between the rock particles. These in turn depend upon the sizes of the particles. The largest particles are rock and gravel. The next smallest ones are sand. The smallest are silt and clay. When silt or clay particles are packed together they let very little water seep downward. Water drains through sand and gravel quite quickly. If it carries clay down with it the clay may fill in the spaces between the sand and gravel a few inches to a foot below the surface. Such a layer may stop more water from passing downward. After heavy rains this waterproof layer may turn the upper part of the soil into a soggy, wet, muddy mess.

The cross section of each type of soil is different because of the differences in thickness of layers and the amount of different-sized particles and decaying material in each layer. For animals or plants that you are investigating you can collect information on the kinds of soils they prefer. You can even collect and preserve samples of the soil profiles for your records and to show others.

It is possible to buy soil augers or soil samplers but they are expensive. You may be able to get a friend who does welding to make you one. You will need about 4 feet (1.2 m) of iron or steel rod and a wood auger bit with a diameter of 1½ inches (3.75 cm) or larger. Cut a foot off the rod and have it welded to form a T with the other piece of the rod. Then weld the shank of the auger to the end of the longer piece of rod.

Where you want to get a soil profile, screw the auger into the ground until the spirals disappear. Then pull straight up. You will bring out the auger with soil trapped in the spirals. Carefully lay the soil on a piece of newspaper and tap it off the bit. Note any layering—color, depth, and the like. Go back into the hole and drill another bit-depth down. Pull it up and tap out the soil just below the previous material. Repeat until you have gone down as far as your auger will permit. You will have revealed a cross section through that soil—that is, a soil profile.

Of course, you can just take a shovel and dig a hole

as deep as possible. This is fine, but it means moving more dirt than you may care to. If you do this keep one edge of the hole as straight sided as possible. Again note all the layers and soil horizons you reveal.

Records of the soil profiles can be preserved in several ways. The simplest is to take a sheet of notebook paper and make a sketch that shows the different layers you have revealed and the depth of each. Make a drawing of the profile to scale. For each layer make a square about 1 inch (2.5 cm) square. Coat the square with rubber cement. Then sprinkle material from your soil sample of that layer onto the rubber cement and let it dry. This will give you a record of that layer's color and texture. For each layer you can also make notes of signs of life you found in that layer such as:

- small insects or spiders
- worm burrows
- plant rootlets
- large animal tunnelways

For shallow soils such as many of those found in woodlands you can make a soil profile by using a shovel or trowel to dig a hole with one straight side. Take a strip from a 4-by-6-inch (10-by-15-cm) file card or an old stiff file folder. Cover it with rubber cement and press it against the straight side of your hole. Let it dry for a few minutes. Remove it carefully. Particles from the soil profile should now be stuck to the cardboard strip.

Similarly, you can use thin strips of wood of greater length. Make them narrower than the size of your soil auger hole. Coat them with rubber cement or contact cement and carefully insert them into the auger hole. Then press it as best you can against the side of the hole. You will get a profile, though not as good a one as if one side were flat. You can protect your soil-profile specimens and get good color by spraying each specimen with clear plastic spray. Don't hold the spray can too close to the specimen or you may air-blast the particles off your board or cardboard. Give each specimen a number and key it to locations on a map. This will give you information to link up to observations you make of wildlife and plants of the area.

You may also want to collect samples of each horizon and put them in separate clear plastic vials. Do not fill any vial more than half-full. Then add water to each and cap tightly. Shake each specimen well and let it settle. The largest particles will settle quickly to the bottom. The different, smaller sizes will settle out more slowly and make layers according to particle size. Use a grease pencil to mark the boundaries between different layers. This will answer your questions about the ability of that horizon to hold water and air. The percentage of different-size particles reveals the relative amounts of spaces that will be available to hold water or air.

PROJECT: MAPPING HABITATS AND VEGETATION

Maps are really just pictures that show where things are. Like any other picture, they can be very sketchy or quite detailed. Often you can start by taking a simple map of your community that shows major roads and waterways and make it more detailed. You might use a light green colored pencil to color in all the places that are open fields and a darker green one to locate the woodlands. Other colors might be used to show marshes and other wetlands. Such maps help point out the places where particular plants and animals might be found.

Maps are not made just of large areas. They can be made of much smaller ones, including just buildings. You might make a map, for example, of your house or school and mark on it all the places you find spiderwebs or wasp nests. On your map you will want to show not only where the animals are located, but the major compass points of north, south, east, and west. Are more of the animals located on a side of the building that faces one of these directions than the others? Are these places wetter, dryer, windier, calmer, hotter, or cooler than the others? Such information can be put on your map and will help answer questions about the relationships between the animals and the habitat.

You might make a habitat map of your neighborhood that shows the major trees, shrubs, and other natural features. On such a map you might plot where you find singing birds such as robins and where you find their nests. You would also mark the places where you see them feeding, drinking, bathing, and the like. As you examine the information you put on the map you will probably discover that they use different parts of the habitat for different activities. Such maps help you answer questions about how different animals interact with their habitat. Different animals use different parts of the same general habitat for different purposes. Just what these are is part of the discovery challenge for you.

Maps help answer questions about how different plants are distributed in an area. You can make a map of a local field or vacant lot and plot on it the locations of all the individuals of a particular kind of plant. Do you see any patterns in how the plants are distributed? On the same map you might plot the distribution of particular soil profiles or moisture conditions. Do you discover any relationships between such conditions and the distribution of the plants? Maybe yes, maybe no. What is the direction of the prevailing wind? If the plant's seeds are wind-distributed you might find that younger plants are springing up downwind of the older ones.

Certain kinds of plants are usually found growing in the same general area as certain other species. Marking the locations of individuals of several different kinds of plants on your maps might indicate patterns of relationships. Just because two kinds of plants are usually found growing together does not mean that they "like each other." It generally means they share the same basic habitat needs. Indeed they may actually be competing with one another for a greater share of the resources of that habitat.

All in all, making maps of the habitats, big and small, that you are investigating is a great way to record information about the animals and the habitats that can help answer some of your questions about how the creatures and their habitats are interrelated and interconnected.

PROJECT: SAMPLING HABITAT CONDITIONS

If the creatures you are investigating are small, like insects, flowers, and many birds and mammals, it is difficult to really study all of their habitats. Instead it is best to take small samples of a habitat and explore them thoroughly. Scientists have developed many mathematical tools to tell how many sample plots should be explored to get reliable and valid information in answer to questions they are asking. At some point you may wish to learn about them. Here we will talk only about some of the ways you might set up a sampling of habitats.

If you are interested in finding out how many of some kinds of plants or animals are in an area, set up some study plots, or quadrats, as they are also known. The size of the quadrat will depend on the size of the creatures you are investigating. For beetles or spiders a square foot may be large enough. For most small things a square yard or meter is best. For trees and shrubs ten square yards or meters is practical.

To make a quadrat take some string a little longer than four times one side of the quadrat. For example, for a square yard cut a string about 4 yards and 2 inches (3.7 meters) long. Tie the ends so that the knot is 1 inch (2.5 cm) from each end of the string. Starting from the knot, mark off 1-yard (.9-m) units by marking the string with a marker.

To set up the quadrat put two stakes in the ground 1 yard (.9 m) apart. Put the knot of the string at one stake, and the first mark on the string should be at the other. Then put two stakes in at the other markers and stretch the string over them. You should now have a 1-square-yard (.8 sq. m) block of habitat outlined by the string.

Another kind of quadrat is circular. To make such a quadrat you will need a piece of old hose or plastic tubing. Select a wooden dowel, or whittle one, that will fit snuggly inside the tubing. To make a circular quadrat about 1 yard (.9 m) in diameter you will need a length of tubing or hose 113 inches (2.8 m)-long and a 2-inch (5 cm)-long dowel. Bend the tubing into a circle and fasten it together by wedging an inch (2.5 cm) of the dowel into each end of the tube.

Choosing the spots for your quadrats is the real trick. Scientists know that the sites should be chosen randomly for the information to be truly useful. That is one of the nice features of the circular quadrats. You can stand at a given point and close your eyes, then toss the quadrat. Where it lands is where you explore. You can make a number of samples by tossing the quadrat from that one point and making your counts of individuals inside the circle. Similarly you can take several colored stones or other easily visible markers. From a chosen point close your eyes and spin around and toss the markers. Where each lands is to be one corner of a quadrat.

Another way to make a sample of a habitat is with a transect. This is a line drawn through a habitat. One way to sample plants with a transect is to set up two stakes at either end of your proposed transect. Tie a string between the stakes. Identify and record all the plants that touch the string or lie directly beneath it.

Transects can be any distance you choose. It is easier to compare information if all transects in a given habitat are the same length. For example you might take two stakes and tie 10 yards (9 m) of string between them. Roll the string up on one of the stakes until you reach the other. Now the transect is easy to carry and to use it all you need to do is stick one stake in the ground, unroll the string, pull it taut, and put the other stake in the ground.

You can combine transect and quadrat. Set up the transect line and choose a unit of distance such as every 2 feet (.6 m). At these points set the corner of a quadrat or the center of a circular quadrat.

The quadrats and transects are just ways of selecting study areas. Inside the areas of a quadrat you can

- count all the individuals of a particular kind of creature
- identify all the different kinds of plants and/or animals you find
- take temperature and other measurements of the habitat
- or make any of a number of different kinds of observations

You will want to compare the findings from different quadrats in the area. Each quadrat will differ in some ways. Some creatures will be more or less abundant than in other quadrats. You may want to average out your findings to get an idea of what is likely to be true of the bigger area of habitat of which your quadrats are only samples.

Using sampling techniques will allow you to get some ideas about whether the plants or animals you want to investigate are usually found in certain parts of their habitats. It can also help you discover if most of them choose the same kinds of places for nests or feeding or other activities. While it clearly is fun to just wander about in nature and see what we can see, curious naturalists often want to get a clearer picture of what is happening. Sampling strategies, although they take a bit more effort, can help get those clearer pictures.

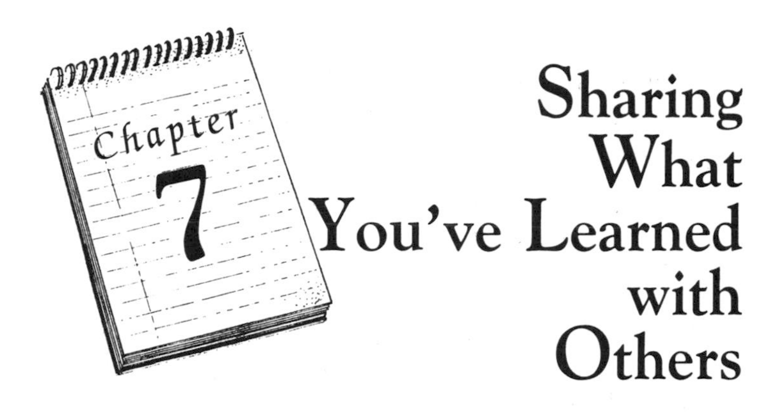

Sharing What You've Learned with Others

Part of the fun of amateur naturalist activity is sharing what you have discovered with others. This may mean just bringing a friend over to show what amazing things you have discovered or collected or it might mean creating an exhibit for school or even a written or oral presentation to your class or some publication such as a local newspaper. It will help in making these communications if you can incorporate photographs or drawings that you have made.

PROJECT: SIMPLE EXHIBITS FOR HOME MUSEUM OR SCHOOL SCIENCE CORNERS

The key to developing a good exhibit lies in simple planning. A good exhibit communicates only a very few big ideas. The first thing to decide is what is the big idea or ideas that you want to share with people. Essentially you are going to tell a story, and the big ideas are the plot of that story.

Big ideas for exhibits based on some of the projects in this book might include:

- Winter is the time ducks are selecting mates on our ponds. Different kinds of local ducks have distinct courting actions.

- We have three major habitats in our community. The major kinds of wildlife in each habitat are . . .

- Milkweed is visited by a number of insects. Many of these share a common color scheme of red and black.

- Different flowers bloom in the fields of our community throughout spring, summer and fall. The blooming season of many of these plants overlap each other.

- Anyone can grow ferns from spores if they follow these steps. Ferns demonstrate an alternation of generations.

- Woolly bear caterpillars relate to the sun. Here's what I found out about that.

Once you have decided on the big story you want to tell, make a list of the key bits of information you will use to tell that story. Be selective. You can't put everything you know in the exhibit because it will become too crowded and people won't take the time to examine it carefully. Next to each piece of information write down how you will communicate that information. It might be just with words but it may be more effectively communicated through a drawing, diagram, photograph, model, or even a hands-on activity.

Figure 7-1 is a sample plan for a project on mallard courtship behavior. The left-hand column describes the information to be presented. The right-hand column gives suggestions for the visual presentation of that information. Figure 7-2 is an outline of the project on growing ferns from spores found in Chapter 5. It lays out the steps involved in growing ferns and also contains notes on visual presentation. You will find such "roadmaps" to be handy tools during the planning stage of a science project.

Figure 7-1; *A Plan for a Project on Mallard Courtship Behavior*

INFORMATION	COMMUNICATION STYLE
The most common waterfowl in our area are mallards and wood ducks.	*Cut out pictures of males and females of each from magazines. Label them with words.*
Mallards and wood ducks are courting from September through March	*Make a small calendar and shade courtship months. Overprint words "Courtship Period"*
Mallard courtship involves group displays and pairing displays	*Write this in words.*
Group displays include Grunt-whistle Down-up Head-up-tail-up Nod-swimming	*For each of the individual displays make a sketch or model diagram to indicate the motions. Write a short word description to explain the diagrams. For help on this I will use the Stokes Nature Guides to bird behavior.*
Pairing displays include Head shake and tail shake Inciting Mock preening Pumping	
Wood duck courtship includes Group displays such as: Bill jerk Whistle-call Inciting	
Pairing displays such as: Inciting Turning the back of the head Display-shake Wing-and-tail flash	
A general thought	*It might be a good idea to include a photograph of a local pond with a group of ducks on it in winter.*

***Figure 7-2.** Outline for Project on Growing Ferns from Spores (see Figure 5-2)*

Steps for growing ferns from spores:

Make a diagram for each stage. Use an arrow to link each diagram.

1. Invert a clean, porous clay pot.

2. Fill the pot with sphagnum, or peat moss.

3. Invert it into a shallow dish or saucer.

4. Pour boiling water over the pot and saucer several times to sterilize them.

5. Drain the water away and cover the pot with a glass jar. Let everything cool.

6. Hold the spore-covered paper over the flower pot and tap the spores off the paper onto the pot and re-cover the pot with its jar.

7. Place the pot and jar in a place with a low to medium intensity of sunlight, or put the setup under a fluorescent lamp, where it can receive eight to sixteen hours of light per day.

8. Keep temperatures between 68 and 86 degrees Fahrenheit (20 to 30 degrees C).

9. Keep the pot moist. Use a mister rather than a watering can. Use distilled water or boiled water that has been cooled.

10. Add a weak solution of liquid fertilizer to the water every couple of weeks after a green mat appears on the pot.

Ferns show alternation of generations

Develop a labeled diagram of each stage in the fern life cycle or include my photographs of each stage.

With such outlines you are ready to develop the actual exhibit. You might find that a large box from the grocery store makes a good "booth" for your exhibit. Cut away the top and front of the box and you have such a booth. It can be spray-painted or covered with contact paper or other background material. Attach the different pieces of the exhibit to the sides or floor of this booth.

PROJECT: SKETCHING AND DRAWING FROM NATURE

Drawings and paintings often capture the eye and raise interest in natural objects and the world of nature. There are many different kinds of nature art. Some is the very technical and precise rendering of natural objects and is generally known as scientific illustration. It is the kind of art that is seen in field guides. Other natural history art is more interpretive. That does not mean that it is not an accurate rendering of the object or objects; it does mean that the artist creates a total picture that communicates something of how the artist feels about the objects. Sometimes that involves the poses of the plants or animals. Or it may be the creation of a sense of mood in the choice of background or even the lack of it.

Some people feel that they just can't draw anything. However, almost anyone can develop skill at drawing from nature with practice and observation. Observation is the heart and soul of good nature drawing. Over time you can train your hand and pencil or brush to make the lines that more and more closely represent what you have seen. The key is to draw and draw and draw. Don't be self-conscious about your sketches or be disappointed that they don't look just the way you think they ought to. Most nature artists keep sketchbooks in which they accumulate hundreds of sketches and partial sketches.

These may be just of paws, or noses, or leaves, or flowers. They are reminders of what they saw in the field. Even what you may think is a poor sketch usually has bits and pieces that reflect some part of the object well.

Later the artist decides what kind of picture he or she wants to develop to communicate some facts or feelings to others. The sketchbooks are brought out to provide reference material as that picture is developed. Photographs from magazines and books may also be used as reference for poses or backgrounds. The finished picture will combine material from many sketches and photos. Whether it is in black and white or color, it will reflect the observations that the person made, especially those which sparked the picture itself.

To communicate with others through drawings and paintings you must develop your skill through endless sketching. Make it a habit to include lots of sketches in your field notebook, along with key words and phrases. If you prefer just to do sketches keep a sketchbook journal. It is useful not to put sketches on the first couple of pages. On these blank pages you will later make a table of contents. Number all the pages after these first blank ones. Then write in what the sketches are on each page in your table of contents. This will help you find reference material later if you want to make a finished drawing or painting to share with others.

There are several good books that will help you build your skills at nature drawing and sketching. You will want to get one or more of these from the library to help you get started and build your skills. Some of these books are listed at the end of this chapter.

You can use your drawings and sketches in many ways to communicate with others. For example, you might:

- Make a set of identification drawings of regular visitors to put by the window near your bird feeder.

- Make a calendar of flower illustrations that show what flowers bloom in your area in that month.
- Make your own nature illustrations for writing paper or cards.
- Make a series of paintings of plants and/or animals that are your wild neighbors and display them in a local library or store front.

Nature art as sharing has played an important role in both building the interest of others in nature and in conserving important resources. It was the paintings of the natural landscapes of the West by Thomas Moran and Alfred Bierstadt that built support from those living in the East for the development of national parks in Yellowstone and Yosemite. Stamps made of wildlife paintings have raised millions of dollars to protect wildlife habitats. Perhaps your own nature art will be able to influence others to an interest in and support for the natural world.

PROJECT: PHOTOGRAPHING NATURE

The camera is a marvelous tool for recording the things you see in nature and giving others a chance to see what they missed. For many people it reveals at least secondhand things they may never see for themselves. Like paintings in the past, photographs in recent years have effectively presented the case for wildland preservation.

The viewfinder of the camera is also a useful observation tool. It helps you focus on a more limited area of any scene. You may even find a fascinating bit of nature surrounded by trash. Through the viewfinder you can block out the trash and frame only the nicer area. However, that does not necessarily mean the camera can record that scene.

Cameras are technical devices and different kinds have different abilities and drawbacks. To make effective photographs you have to learn what these are for the type of camera you have available to you. The camera you have may not allow you to make the kind of photographs you would like.

Many easily available cameras are fixed focus. That is, objects within a certain range of distances from the camera will be in focus on the film. Others will be fuzzy and out of focus. For many of these cameras you cannot get clear pictures of objects closer than 3 feet (.9 m). These cameras are very good for taking pictures of landscapes and large animals but not for taking close-up pictures of smaller things. Some of these cameras will permit the use of snap-on "portrait lenses" that will let you get somewhat closer and still be in focus.

To take good close-up photographs or to get large images of distant objects you will need access to a camera with interchangeable lenses. Then you can choose and use a lens that can focus at the distances you find the objects you want to photograph. Most of these cameras are single-lens reflex 35-mm cameras. Unfortunately, both the camera body and the lenses of these cameras are expensive. However, if you want to be serious about nature photography, this is the type of camera you will need for most of your work.

Today more and more families have videocameras. These can be very useful for nature work because they not only capture moments in time but whole sequences. Thus, with the motion you can show activities and processes not possible with the still cameras. You will be able to capture such things as a butterfly uncoiling its proboscis to sip nectar or spring peepers blowing up their "bubble gum throats" and uttering their shrill peep. You can record the tail-signaling language of squirrels or the courtship behavior of ducks on a pond.

Like wildlife sketching, still or motion photography

takes practice to develop skill. There are many choices available in terms of cameras, films, lenses, and other accessory equipment. In this book we cannot begin to go into the detail of any of these choices. The key is to determine what equipment is available to you and learn all you can about it. Determine what its strengths and drawbacks are. Do all you can with those strengths. In time you may be able to get access to other equipment that overcomes the drawbacks.

Whatever the photography equipment you have, you can use it effectively to communicate your investigations and feelings about the natural world to others.

PROJECT: COMPUTER NETWORKING

Computers have opened up new ways of sharing your findings about the natural world with others of similar interest. Many more families today have home computers and a number of these also have modems, which permit them to use the phone lines to connect to thousands of other computers around the world.

There are now programs available for keeping birding field notes on the computer, and these can be shared with others using the modem. Your field notes can be kept on spreadsheet programs or on a MacIntosh program called Hypercard. These, too, can be shared with others by using a modem. Day by day new ways, with which naturalists can use the computer to share information and thoughts with others, are being developed.

National Geographic has developed a project called KidNet. This project involves young people in collecting scientific information and sharing it with other young people around the nation. Scientists work with this information and periodically report on the network about the overall findings of those who participate.

Another organization, called TERC, is doing something similar with students from around the world. These young naturalists and scientists are gathering and sharing information on the sequences of natural events, such as the blooming calendar mentioned earlier.

You can also get involved with a number of electronic bulletin boards and electronic mail if you have the computer and modem. These items let you communicate with groups of people or other individuals with similar interests around the country at fairly low cost. It is a great deal of fun to make those connections and share interests and findings with other people. Some detailed information on some networking opportunities follows.

SOME REFERENCES FOR DEVELOPING SKILLS FOR SHARING WITH OTHERS

Leslie, Clare Walker. *Nature Drawing: A Tool for Learning*. Englewood Cliffs, N.J.: Prentice-Hall, 1980.

Leslie, Clare Walker. *Field Sketching*. Englewood Cliffs, N.J.: Prentice-Hall, 1983.

Rue, Leonard Lee, III. *How I Photograph Wildlife and Nature*. New York: W. W. Norton and Co., 1984.

West, Keith. *How To Draw Plants: The Techniques of Biological Illustration*. New York: Watson-Guptill Publishing, 1983.

COMPUTER NETWORKING

National Geographic Society Kids Network
National Geographic Society Educational Services
17th and M Streets NW
Washington, DC 20036
Phone: 202-857-7000

(Inquire about how you can share your findings with other young people around the country using the computer.)

EcoNet
18 DeBoom St.
San Francisco, CA 94107
Phone: 415-442-0220

(Inquire about how you can use the resources of EcoNet to find out about many aspects of the environment and particularly how you can share information with other young people around the globe using I*EARN, sponsored by the International Education and Resources Network)

Global Education Project
TERC
2067 Massachusetts Avenue
Cambridge, MA 02140
Phone: 617-547-0430

(Inquire how you can participate in any of the phenology investigations being conducted by young people around the globe. Phenology is the investigation of the recurring events in nature, such as the first and last arrivals of migrants, the blooming of plants, and the like.)

Other Ideas for Amateur Nature Projects

In earlier chapters of this book we have suggested a few projects to be explored by the amateur naturalist, and some strategies for exploring the natural world. Here we suggest many more, along with some tips and tricks on how to get started with them. These are only meant to challenge your own curiosity. As you get involved with any of them you will develop your own questions and figure out ways to get the answers to them.

The earth is a place that is undergoing constant change. Plants and animals adapt to these changes. The same species may respond to similar situations differently in different places. The information you find in books about nature may be true for a particular place and time but may not apply to the plants or animals where you are. For example, in most of the northern states, where woodchucks live, they hibernate beginning in the fall. However, in the southern part of their range some woodchucks may not hibernate at all. Birds—like crows, for example—develop local dialects to their calls. Individuals from one region may not recognize the calls of the same species from elsewhere.

The curious naturalist uses and enjoys good nature books but still spends much time investigating nature itself. From many observations over time, he or she may find local creatures behaving somewhat differently from what is described in the books. Become an active hunter of information about the lives of your plant and animal neighbors.

KEEP A MIGRATION CALENDAR

Many species of birds migrate long distances between their wintering grounds and their nesting grounds. A few species spend all their lives in the same place summer and winter. You can have fun bird-watching and learn a lot by keeping a bird calendar.

Choose a calendar with large open blocks for each day. Each day that you see a particular species of bird, enter its name on your calendar. You may also want to enter a number for how many individuals of that species you saw that day (Figure 8-1).

Some species are indicators of changing seasons. New Englanders are interested in such things as when the first redwing blackbirds, robins, or bluebirds return, for this is an indication of the incipient arrival of spring. They also look for the arrival of flocks of shorebirds, such as semi-palmated plovers, for this indicates that autumn is fast approaching. The arrival of flocks of juncoes suggests that wintery weather will be arriving soon from the north.

You may also get to see the occasional wanderer, a species that gets blown off its normal migration route by storms and ends up in unusual places. Your calendar may also show the late-summer arrival of some of the heron family that usually are found breeding in the south. When their young can fly and set out on their own they often move northward to new areas. All in all, keeping a migration calendar of birds can be a fun and enlightening activity for years to come.

CONDUCT A CENSUS OF BREEDING MALE BIRDS IN YOUR AREA

In spring, male songbirds claim breeding territories and announce their presence to potential mates and rivals

Figure 8–1. Two types of bird-sighting logs

DAILY SIGHTING LOG

MONTH: June YEAR: '93 LOCATION: Yard

	1	2	3	4	5	6	7	8	9	10	11	12	13	14	15	16	17	18	19	20	21	22	23	24	25	26	27	28	29	30	31
Mourning Dove	✓	✓	✓	✓	✓	✓	✓	✓	✓	✓	✓	✓	✓	✓	✓	✓	✓	✓	✓	✓	✓	✓	✓	✓	✓	✓	✓	✓	✓		
Black-capped Chickadee			✓				✓	✓	✓			✓	✓	✓	✓			✓		✓		✓	✓			✓	✓		✓		
Horned Lark			✓				✓	✓	✓				✓	✓		✓	✓	✓		✓		✓		✓	✓	✓		✓	✓		
Meadow Lark	✓	✓		✓	✓	✓		✓		✓	✓	✓	✓	✓		✓	✓		✓	✓	✓			✓	✓	✓	✓	✓	✓		
Crow	✓				✓	✓	✓		✓	✓	✓	✓	✓	✓	✓	✓		✓	✓		✓	✓	✓	✓	✓		✓	✓	✓	✓	
Killdeer	✓		✓	✓	✓	✓		✓	✓	✓	✓	✓		✓	✓	✓	✓	✓		✓	✓	✓		✓	✓	✓		✓	✓	✓	
Bluebird			✓								✓	✓	✓										✓					✓		✓	
Downy Woodpecker						✓						✓										✓				✓					
Goldfinch	✓								✓	✓					✓	✓			✓					✓			✓				
Mallard															✓	✓		✓													
Blue Jay	✓		✓		✓	✓	✓	✓			✓	✓			✓	✓			✓		✓	✓	✓	✓		✓		✓	✓		
House Sparrow	✓		✓	✓	✓		✓	✓	✓		✓	✓	✓		✓	✓		✓	✓		✓	✓	✓		✓	✓	✓	✓	✓	✓	
Barn Swallow							✓		✓						✓					✓	✓				✓			✓		✓	

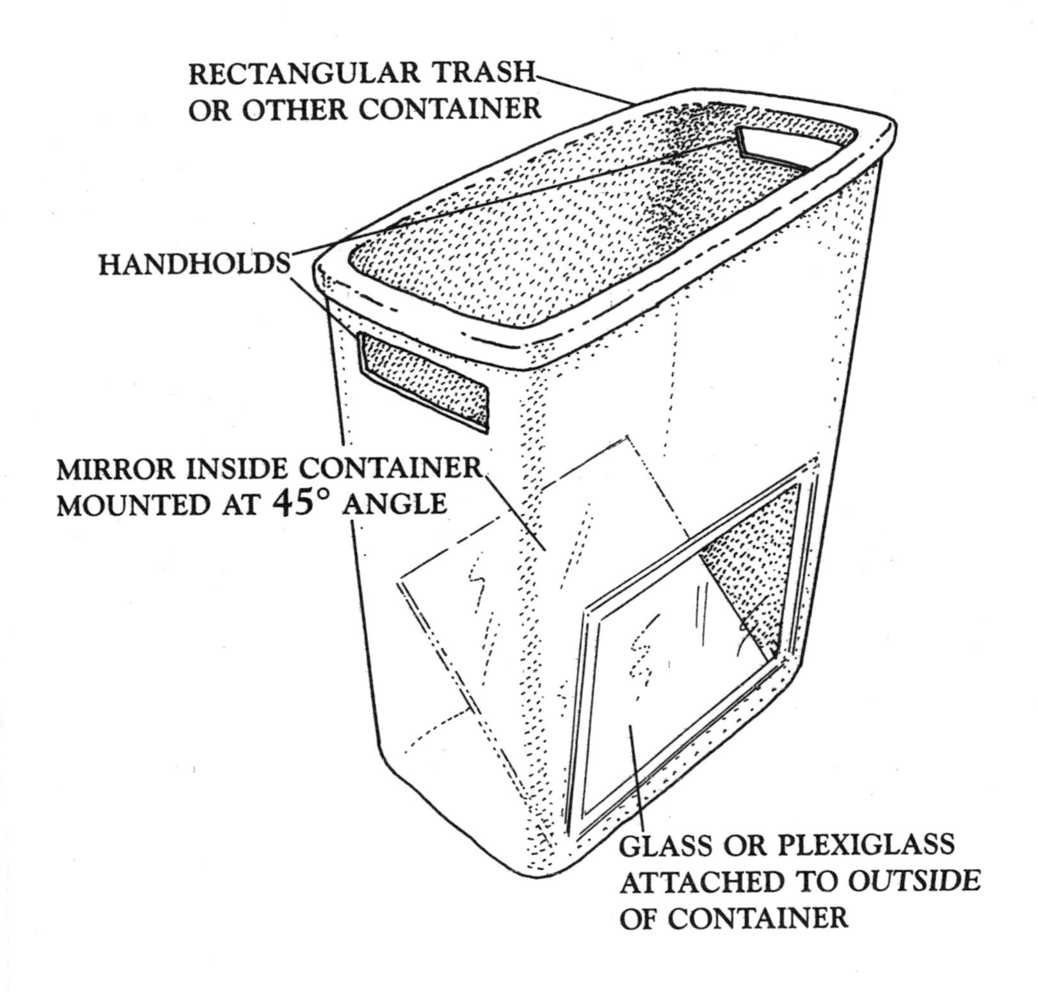

Figure 8–2. *A simple waterscope you can make to see underwater.*

alike by singing. There are a number of recordings available to help you identify the birds by their songs. You can use them to train yourself to recognize the songs of the most common birds of your area.

Pick a route through a good local bird habitat that you can travel regularly on foot. Or one that you travel on bicycle. Stop at regular intervals and listen for and write down the time and location of each bird of each species heard from that spot.

Follow your route and stop at each of your listening points at least two or three days a week if possible. You will get some good information even if you are limited to following your route only on weekends. Early morning and late evening are the times when most singing is done. Plan your trips for these times if possible.

The information you gather will help you know how many adult males of each species live in your area and where they have set up territories. Not all will find a mate. Some will keep singing well into the spring after most of the others have become too busy with family matters to still be singing. Most of these late singers have not found a mate.

The ones that have stopped singing regularly have probably found mates and are raising a family. A few may have stopped because they became a meal for some predator. To determine which, you will have to look about and see if you find individuals of that species in the silent singers' territories.

EXPLORE CATERPILLAR FOOD PREFERENCES AND FIND OUT WHICH CATERPILLARS BECOME WHICH BUTTERFLIES OR MOTHS

Caterpillars are fascinating little hydraulically driven eating machines. Caterpillars, as they grow and shed their skins, require ever greater amounts of plant food. A group of caterpillars on a plant can destroy a great many of the plant's leaves. They may even strip it of leaves completely.

Plants fight back chemically. They make chemicals that may be distasteful or even unhealthy to particular types of caterpillars. Over time some kinds of insects develop ways to counteract the chemicals. The plants then must develop new ones. The end results are that some plants can be safely eaten by one kind of caterpillar but not by another.

You can keep and feed caterpillars through their several molts and watch them make chrysalids or cocoons depending upon whether they are butterfly or moth caterpillars. In time, if all goes well, you may observe the adult butterfly or moth emerge. While following these life histories you can also learn something about what plants the caterpillars will and will not eat.

BUILD A WATERSCOPE TO EXPLORE UNDERWATER LIFE IN A POND, STREAM, OR TIDEPOOL

There is a whole fascinating world of creatures that spend their lives in water. There are several ways we can spend time watching them go about their lives. We can catch them and put them in an aquarium. (More about that shortly.) We can put on a face mask and snorkel and join them in their world for a while. Or we can make a waterscope so that we can wade or sit in an area and see clearly underwater.

The problem of course is that the water's surface tends to reflect light and make it difficult to see beneath the surface. There are polarized glasses that you can wear that will reduce the reflection, but a waterscope removes the reflection problem completely. With a waterscope you can look directly below you from a boat or you can use it wading about in the shallows.

Perhaps the simplest waterscope you can make is with a plastic bucket, some bathtub silastic sealant, and some glass or plexiglass (Figure 8-2). The plexiglass is easier to cut and less apt to break. However, it will scratch easily from sand and gravel on the bottoms of ponds and streams.

Carefully cut a hole in the bottom of the bucket. If you are going to use glass the hole should be square or rectangular. If you use plexiglass it can be round. Be sure that there is at least 1 inch (2.5 cm) of space between the edge of the bucket and the edge of the hole you cut.

Cut the glass or plexiglass so that it is ¾ to 1 inch (2 to 2.5 cm) larger than the hole you cut in the bucket. Spread sealant all around the edges of the hole on the *outside* of the bucket. Press your glass or plexiglass onto the sealant. Be sure that there is a good solid seal with no air bubbles. If you were to put the glass or plexi on the inside of the bucket, water pressure would push it away from the sealant.

At opposite sides of the rim of the bucket and about 1 inch (2.5 cm) below the rim, cut slots. The slots should be rectangles about 1½ inch (3.75 cm) deep and a little wider than the width of your hand. These make hand holds so you can more easily control the waterscope when you are pressing it into the water.

You will be able to see better if you spray-paint the inside of your bucket with flat black. Put something over the glass on the bottom before you start spraying, or paint it before you attach the glass or plexi.

With such a waterscope you will be able to see water life directly below you or at small angles in any direction.

A waterscope that gives you a broader field of view

can be made with a bit more work. The basic principle is the same, however. Instead of a round container you will need a rectangular one. This can be built from waterproof plywood or you can use a rectangular plastic waste basket. For this waterscope you will use a rectangular window cut in the side of the container about an inch (2.5 cm) from the bottom. The viewing glass or plexi would be attached as above.

The big difference in this scope is that you mount a mirror inside the container running from the opening to the back wall at a 45-degree angle. This allows you to look down but see what is in front of you. You may want to put a cover on top of the scope with just a small space through which you look down to the mirror. This will block out unwanted light from above. You can also make a cover that has a hole just the size of your camera lens. You can then take pictures of what you are seeing underwater.

With this type of waterscope you can see much more of what is going on underwater. You can also adjust things so you see above and below the water at the same time.

Waterscopes can open whole new worlds for you both in freshwater and in tidepools at the seashore.

KEEPING TRACK OF LOCAL TURTLES BY MARKING THEM AND RELEASING THEM

If you explore waterways in your area you often see turtles sunning themselves on rocks and logs. Do you ever wonder if you are seeing the same turtles each time? Or how far do individual turtles move around in this area? Or is this the same turtle I saw here last year? The problem is that individual turtles look pretty much alike except perhaps for size.

To recognize a number of individuals you have to

mark them in some way. We would of course want to mark them in a way that will last but is not harmful to their health or painful to them. There is a way (Figure 8-3).

The shell of a turtle is made up of bone covered with thin plates of material that is much like your fingernails. Each plate is called a scale or scute. The overall shell has two major parts. The back shell is called the carapace; the belly one, the plastron. There are a number of scutes around the edge of the carapace. Where the carapace goes over the hind legs and tail, the scutes and carapace make a thin shelf. This continues to the point where the plastron and carapace are joined. This area is divided into equal segments, usually five, by the marginal shield of scutes.

Turtles are marked by cutting notches into, or drilling holes with a hand drill through, the marginal shield scutes in a pattern. Drilling is actually better because nothing in nature will cause such a mark. However, raccoon teeth, or those of other predators, may make notches on the edge like those you might cut. In drilling, care must be taken not to nick the leg in the process. That is why a hand drill rather than a cordless electric drill should be used.

Another advantage of a drilled hole is that colored pipe cleaners or wires can be twisted in the holes of some individuals to make them more recognizable at a distance.

Take a turtle and with the animal facing away from you notice that the marginal shield scutes generally are five on either side of the midline of the carapace above the tail. There are five to the right and five to the left. Start with one at the back and five to the front. If you notch or drill the scute to the right of the midline this is R1. The next turtle you mark might be R2 and so on to R5. Then start with L1. When you have marked ten turtles the next might be R1, L1. There are enough

Figure 8–3. *Identify members of your local turtle community by drilling holes in the marginal shield scutes.*

combinations for you to individually mark most of the turtles in your local pond. If you have lots of turtles and run through the combinations, move to two holes per scute. This would be recorded as 2R1 and so on.

Of course, unless you are using the wire flags, you must usually recapture the turtle to read its marking. Turtles can be caught by net from boats or canoes or they can be caught in "sunning traps." Sunning traps are made by constructing a rectangular form of wood or Styrofoam. Attach to this a cage of hardware cloth or fish netting. Around the inside of the frame put finishing nails about an inch (2.5 cm) apart. You may also want to attach a string and rock to the trap to anchor it in one place.

Turtles will climb onto the float to sun and will enter the inner pool either by accident or for bait you may place there. The nails will prevent the turtles from climbing back out until you remove them. Remember that it takes energy for even a turtle to swim about. Any traps should be checked every day and the trapped turtle removed, marked or checked, and released. If the traps are not checked the turtles can drown. If you are not going to be able to check a trap for several days, take it out of the water during those times.

COLLECT WILDLIFE CLUES

Where wildlife lives you will find a variety of clues to their presence. These may include shed hair, feathers, or skin. Also the waste droppings of animals make important clues not only to their presence but to what they are feeding upon. Some animals, such as squirrels, leave middens of pinecone scales and cobs, partly opened nuts, and other food debris. Abandoned nests of birds and mammals also reveal much about the habits of their makers.

Clearly you won't want to collect all the clues you find but it is useful to collect samples of each to serve as a reminder and reference at a later date. Collections of such clues are like the seed collections talked about earlier.

Mammals shed their fur periodically. Fur also often clings to bark or twigs when the animal brushes by or squeezes into a den. Sometimes the fur will be found on barbed wire or on burrs the animal has scratched out of its coat. There are no simple identification books to hairs. The best way to find out about hair clues is to compare them with a reference collection you have made.

While it is not wise to handle road kills, it is not too risky to pull out a few long guard hairs and some belly fur. Put this in clear plastic vials and insert a slip of paper with the name of the animal the hair came from. You can also ask hunters and trappers in your area to collect samples of mammal fur for you.

As you build your collection and observe it, notice the pattern of banding of colors on the hair. The width, sequence, and color of each band is often typical of a particular species. If you have a microscope, examine the hairs with a high power. You will see scales along the hair. The distinctive patterns of the scales are also often specific to a species.

When larger animals walk on soft ground they leave tracks. On dusty surfaces even small creatures will leave tracks. These tracks can be collected by photography or by making casts of the tracks. Casts are most easily made of tracks in mud and soft ground.

To make a cast of most tracks you will need plaster of paris or the finer powdered dental casting plaster (Figure 8-4). You will also need water, a mixing container, a stirring stick, and some grease.

Once you have chosen the track to cast you will need to make a collar around it so that the plaster will

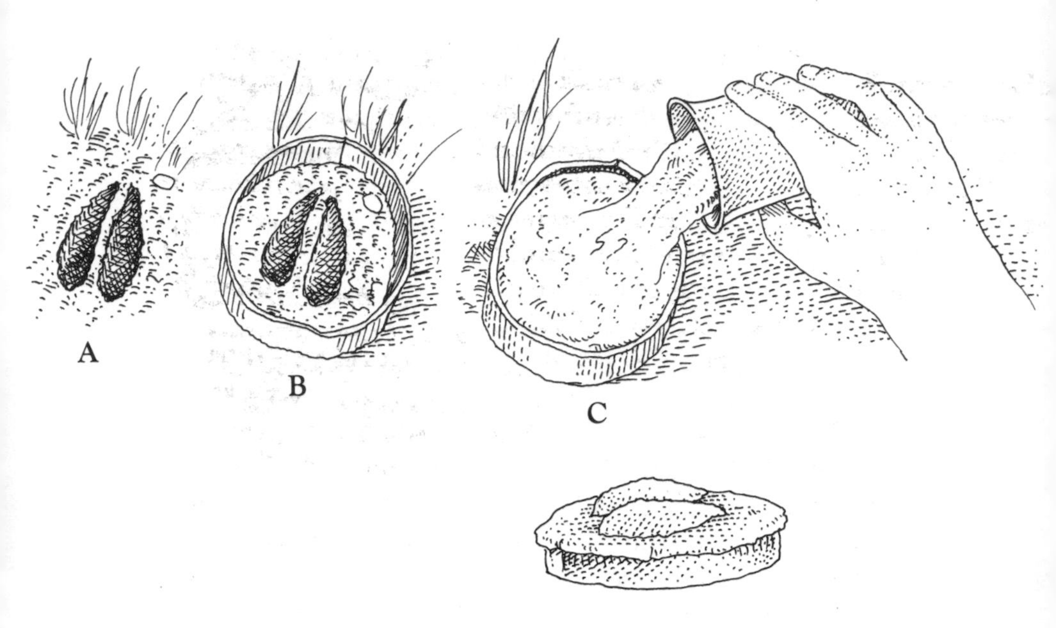

Figure 8–4. *A simple way to make casts of animal tracks using cardboard strips and plaster*

not run all over the place. A collar can be made of cardboard strips about $1\frac{1}{2}$ inches (3.75 cm) wide and long enough to encircle the track with an inch or two (2.5 to 5 cm) to spare in all directions. For many small animal and bird tracks, a cat-food can with both ends removed makes a dandy and sturdy collar. The inside of the collar should be well greased so that the plaster will not stick to it.

Mix the plaster with water in the container. Old plastic jugs are good containers for this. The plaster-and-water mix should be about the consistency of thick cream. It should flow slowly, not plop. Pour the mix into the collar to a depth of at least 1 inch (2.5 cm).

Let it set for an hour or so until it is thoroughly hardened. Then remove the hardened plaster from the collar and brush or rinse away the dirt that clings to it. You will now have a positive cast of the track. This you may want to paint to make it stand out from the white of the plaster.

You may also want to make a negative cast of the track that will look indented like the original track on the ground. Put a collar around your cast with the track facing up. Spray the collar and cast with spray cooking oil. The spray will help ensure that all the nooks and crannies of the cast are covered. Then pour a new mix of plaster into the collar and let it set. If the original was well greased, the two pieces of plaster should separate easily. You will now have both a negative and a positive cast of the track.

Casting a track in snow is more difficult and not always possible. It depends a great deal on the outside temperature and the type of snow. It is almost impossible to cast a track in powder snow. If the snow is wet and sticky the chances of making a successful cast are better.

When plaster and water are mixed they give off heat. This means that when the plaster is poured it is likely to melt the track and distort its size and shape. There are two tricks you can use to try and overcome this problem. The first is to carry a plant mister with you inside your coat when you set out to cast your tracks. When you find a track, very gently spray a fine mist over it. If conditions are right, a thin film of ice will form over the surface of the track and its surroundings. Quickly mix the plaster and pour. If your luck holds, the plaster will set before the ice film melts.

If the weather is a bit warmer and water mist is not likely to freeze quickly to ice you can try using a can of clear plastic spray to make the protective film. Don't hold the spray can too close to the track or the force of the spray will change the shape of the track. Only experience and practice will give you good snow-track castings.

The best way to preserve a snow track is with a camera. Best results can be had when the sun is relatively low in the sky, morning or evening, so that shadows give shape and dimension to the track. Correct exposure of film on snow is difficult. There is often more light reflecting into the lens than the light meter indicates. This means the picture will be overexposed. If you have a camera on which you can change the f-stops, take extra pictures of one or two f-stops smaller (larger number) than you would normally use. With experience you will learn how much smaller the lens opening should be for your snow conditions.

If you use color film be sure to use a skylight filter or a filter that removes extra blue tones in the shadows. A polarizing lens is also a useful supplementary lens to reduce or eliminate snow glare.

If neither film nor casts are possible for you, simply make careful drawings of the tracks complete with measurements of length and width of each individual track and measurements of the distance between tracks. The distance between tracks is the stride and it will help you tell if the animal is walking or running. Each action of the animal produces a slightly different pattern of the individual tracks. A sampling of different tracks can be found in Figure 8-5.

When following tracks it is useful to carry a stride stick. This is simply a straight stick that you can use to mark with a pencil or chalk the normal stride of the animal you are following. If you become confused following a set of tracks, because the next one seems to have disappeared, use the stride stick. To do this lay the stick in front of the last track you can clearly see. Note where the next track should fall according to the measurement of the stride. Scan the ground at this dis-

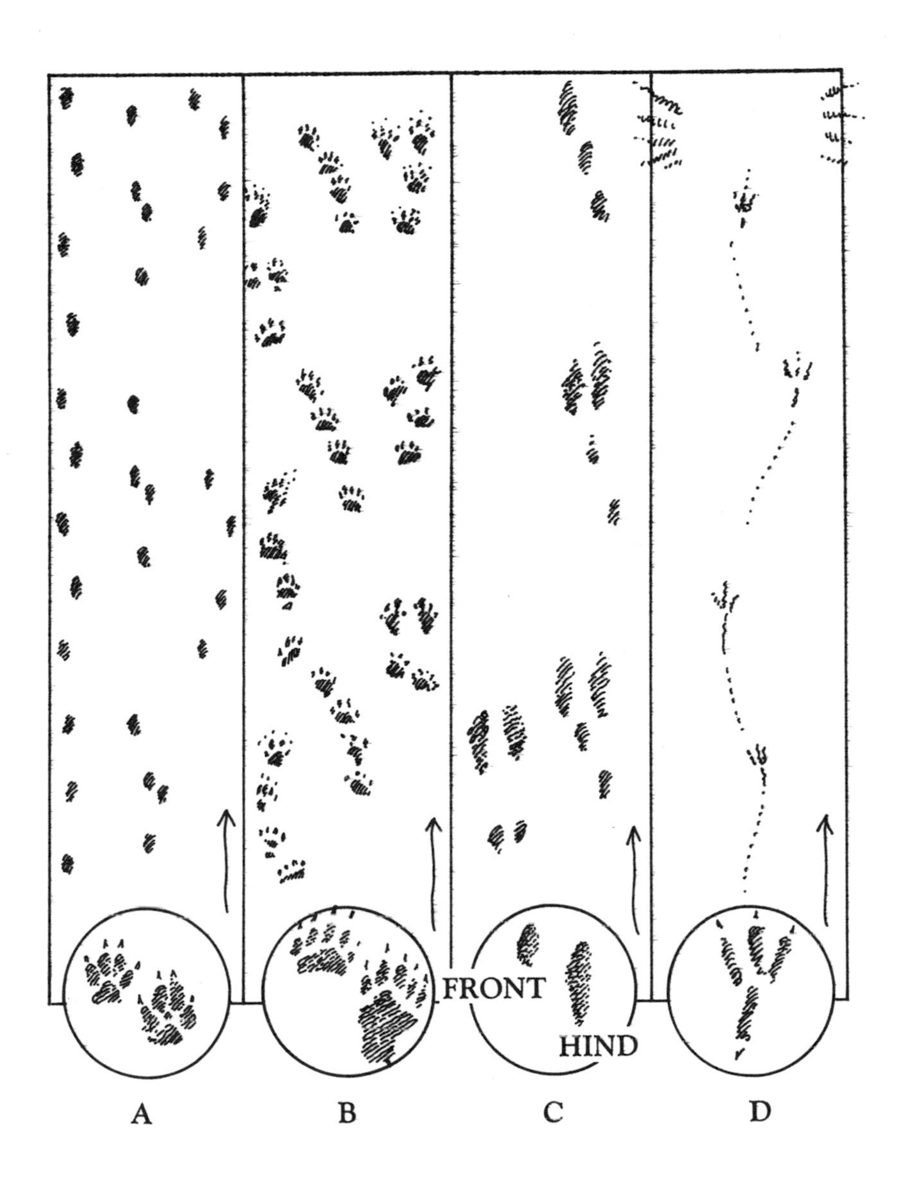

Figure 8–5. A *sampling of animal tracks:*
A. *red fox,* B. *skunk,* C. *cottontail rabbit,* D. *crow*

tance. You often will be able to find some faint evidence of the "lost" track. If not, put a marker where you think the track should be according to the stride stick. Move the stride stick one stride further along from the marker. You may pick up the next track now.

EXPLORE AGING CLUES

All living things grow over time. There are a variety of different clues left behind that help the naturalist tell the approximate age of objects or events. For example, with humans, we know that they lose their first set of front teeth between five and six years old and that they get a major set of molar teeth at around twelve years old. A person with or without these teeth is within certain age ranges. Trees grow a layer of open porous wood in summer and a thinner, denser layer of wood in winter. These two layers combined form the growth of one year. If the tree is cut across with a saw these layers appear as a ring in the cross section. The dense winter layer is seen as the boundary of each annual ring. By counting these rings it is possible to make a close estimate of how old the tree was when it was cut.

The key to aging an individual in both these examples is knowledge of its kinds of patterns of growth. Annual growth spurts that create ring patterns similar to that of trees can be found in the individual scale patterns of fish and some turtle shells. Many mammals demonstrate specific timing for emergence of different teeth just as we humans do. Furthermore, as the mammal goes about the process of living its adult teeth wear down. By examining the wear on the teeth of a mammal skull it is possible to tell about how old the animal was when it died. The feather colors and patterns of color of most birds differ from juvenile to adult. All of these features, once known in detail, can help someone determine the approximate age of an individual of a particular species.

Most cone-bearing evergreen trees, such as pines, grow from a whorl of buds at the tip of a branch. The center bud produces the leader that grows straight up or out. The other buds form the side branches. The length of the leader tells how much the tree or branch grew in one year. Technically, one year's growth is a twig. If you look at a young pine tree and count the number of whorls of branches around the trunk, you have a good approximation of how old the tree is. Measure the distance between whorls and you know how much new growth occurred that year.

In making age and growth determinations on coniferous trees start from the top down. As the tree ages, its lower branches get shaded and die. The bark eventually grows over the stubs and they disappear. This of course affects the accuracy of your age estimate. There are other needs for aging than knowing the absolute age of the tree. For example, many pines have their leader invaded and killed by insects. In such cases, one of the side buds will turn upward and take over the role of leader. You can tell how many years ago, and thus in what year, that particular event took place on that tree. You must, however, count down from the top of the tree.

Hardwood trees do not have the growth pattern of the pines. Aging by branch whorls doesn't work. However, cut hardwoods often send up sprouts from cut stumps. Almost no conifers do. There are clues on these sprouts to help you determine when the main tree was cut. The cutting might be done by ax or saw or it might be by beavers. I have frequently used the method I'm about to describe to determine whether or not beaver activity in an area is recent.

Buds, which house a tree's leaves and flowers, are

covered with scales. When these scales fall off, as the buds open they leave a scar on the branch. Bud scales on the sides of branches leave only short scars. Those from the bud at the tip of the twig leave scars all around the twig. The distance between these completely encircling scars is one year's growth (Figure 8-6). By locating each of these encircling scars back along a branch you can determine approximately how old it is.

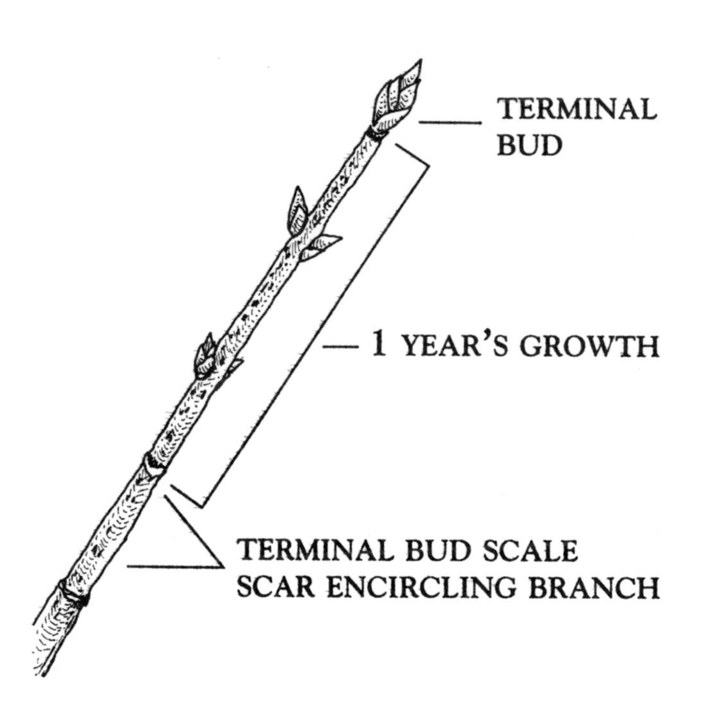

Figure 8–6. *The distance between two terminal bud scars on a tree branch indicates one year's growth.*

Begin at the tip of the branch and trace back until you find the encircling scars. This will be a complete encircling in the case of species with a true terminal bud; it will be almost complete in those species with an offset end bud. That distance will represent the latest growing season. Continue back to each succeeding set of encircling scars. You can usually trace back between three and five years. Beyond that, new bark tends to obscure the scars. The distance between these scars indicates how much growth occurred for that year.

The amount of annual growth will vary by species and by local environmental conditions. Even different branches on the same tree will show different amounts of growth in the same year. It is fun to trace the growing history of a particular tree for its more recent years.

When hardwoods are cut they often send up sprouts from buds beneath the bark at the base of the tree. Aging these stump sprouts will tell you approximately how long ago the parent tree was cut. What you cannot tell is if the tree was cut in spring or fall. If it was cut in spring, it may have sent the stump sprout up that growing season. If cut in late summer or fall, no sprout will start until the next growing season. Thus if you age such a sprout at three years, you must say that the parent was cut between three and four years ago. It helps to age a number of such sprouts in a particular area and get an average age for cuttings. This is what is done in assessing beaver activity around a beaver colony.

EXPLORING THE MILKWEED COMMUNITY

Milkweeds are widely distributed plants found in open lands around much of America (Figure 8-7). They are an interesting group of species. They are characterized by sticky milky sap and bulky seed pods that produce seeds with fluffy tufts that carry them airborne to new

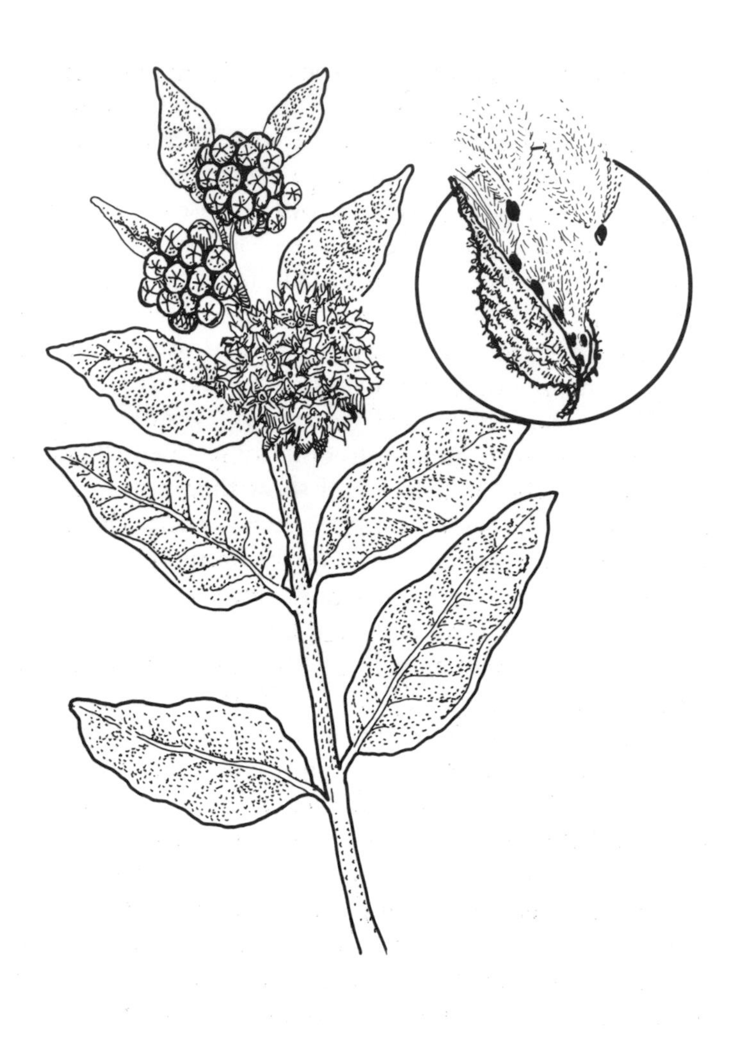

Figure 8–7. *The milkweed plant is perhaps best known in its role as home to the monarch butterfly.*

locations. Milkweeds also reproduce by underground runners so that once established, a single plant soon becomes a whole colony.

These colonies of milkweeds are the home community to a number of different groups of insects. These insects use the plants in different ways but often share some common characteristics such as a red-and-black color scheme and an unpleasant taste to predators. Probably the best known of the cohort of milkweed dwellers is the monarch butterfly, renowned for its migrations and hibernation activity.

Milkweeds are a good place for amateur naturalists to explore the variety of relationships and interrelationships between and among different species. Explore such questions as:

- Where on the plants do most of the monarchs lay their jewel-like eggs?
- Are milkweed bugs and milkweed beetles more active at certain times of the day than others?
- How much do the bugs and beetles move around among individual plants? To determine this you may want to individually mark the insects with tiny drops of nail polish or airplane dope. Keep the drops small so that the solvents in these products do not harm the insects. You can create a code similar to that used in marking turtles by thinking of the wings of the insects as being right or left and arbitrarily divided into front, middle, and rear sections.

Take time to examine the milkweed flowers with a hand lens. Can you detect how the flowers actually trap some insects such as bees to increase the likelihood of their getting covered with pollen? The more you explore this interesting group of plants and their associated animal life, the more you will find to pique your curiousity.

MAKING SPORE PRINTS OF MUSHROOMS

Mushrooms were once considered to be a type of plant, one that lacked the characteristic green material chlorophyll that other plants possessed. It was known that they didn't need chlorophyll to make food because they got their food from the decay process. Today mushrooms and their allies are classified not as plants but in their own separate kingdom—the Fungi.

Fungi are creatures composed of a web of threadlike cells called a mycelium. The mushrooms that we recognize are only the reproductive part of the individual fungus. The various threadlike structures, called hyphae, fuse together to form the characteristic mushroom, and inside the mushroom on pores or gills they develop the reproductive spores. Spores vary in color from species to species. Often mushrooms of different species will look very similar but their spores will differ in color. These differences help us identify the different kinds. Four or five main groups of colors occur. To determine the colors of the spores, we can make what are known as spore prints.

To make a spore print, remove the mushroom stem and place the cap with the gills down on a sheet of white paper or glass. Cover the cap with an inverted cup or glass to prevent breezes from disturbing the spores. Let the cap remain undisturbed for at least six to twelve hours, preferably twenty-four hours. Then, when you carefully remove the mushroom cap, spores should be deposited on the paper or glass as a print. You can preserve the print by letting clear plastic spray settle onto the print. If you spray the print too directly it is likely to smear or otherwise destroy the pattern.

The spore colors will fall into one of the following categories: white, pink, brown, or blackish violet. This last category is often divided in two: black and purple. A few rare species have green spores. The color of the spores is one of the criteria for identifying species of fungi. No one criteria is sufficient. You must use a combination of criteria to be sure of the identification of any mushroom.

Some people make very handsome art pieces by arranging spore prints on paper. You may even use different-colored papers such as using a dark blue as background for mushrooms that leave a white spore print. You might even use mushrooms that leave different-colored spore prints to make a collage of such prints. Use your imagination and creativity.

CREATE A FROG CHORUS CALENDAR

Most naturalists are attracted to water. All animals need water and wetlands are often places where wildlife concentrates. Among the fascinating wildlife of waterways are the frogs (Figure 8-8). Many naturalists get their start by becoming acquainted with the various frog choruses in their area.

In the Northeast, the first frog to start the chorus is the wood frog. The first males to the recently melted woodland pools float in open water. There they inflate the sacs on the sides of their heads and start their quacking calls. Other males come from the surrounding forest to join them until a full chorus is sending its "come hither" sound throughout the area. Females follow the sound to its source. The mating rituals follow.

Wood frog choruses are soon joined by spring peepers, usually from areas more marshy than wood frogs prefer. Hidden beneath an overhanging leaf, the males inflate their throats as if they were made of bubble gum and make their extraordinarily loud whistled peeping note. Peepers are hard to spot at first. A flashlight beam reflecting off the inflated throat is usually the first glimpse. Once you have located one, it becomes easier and easier to spot others.

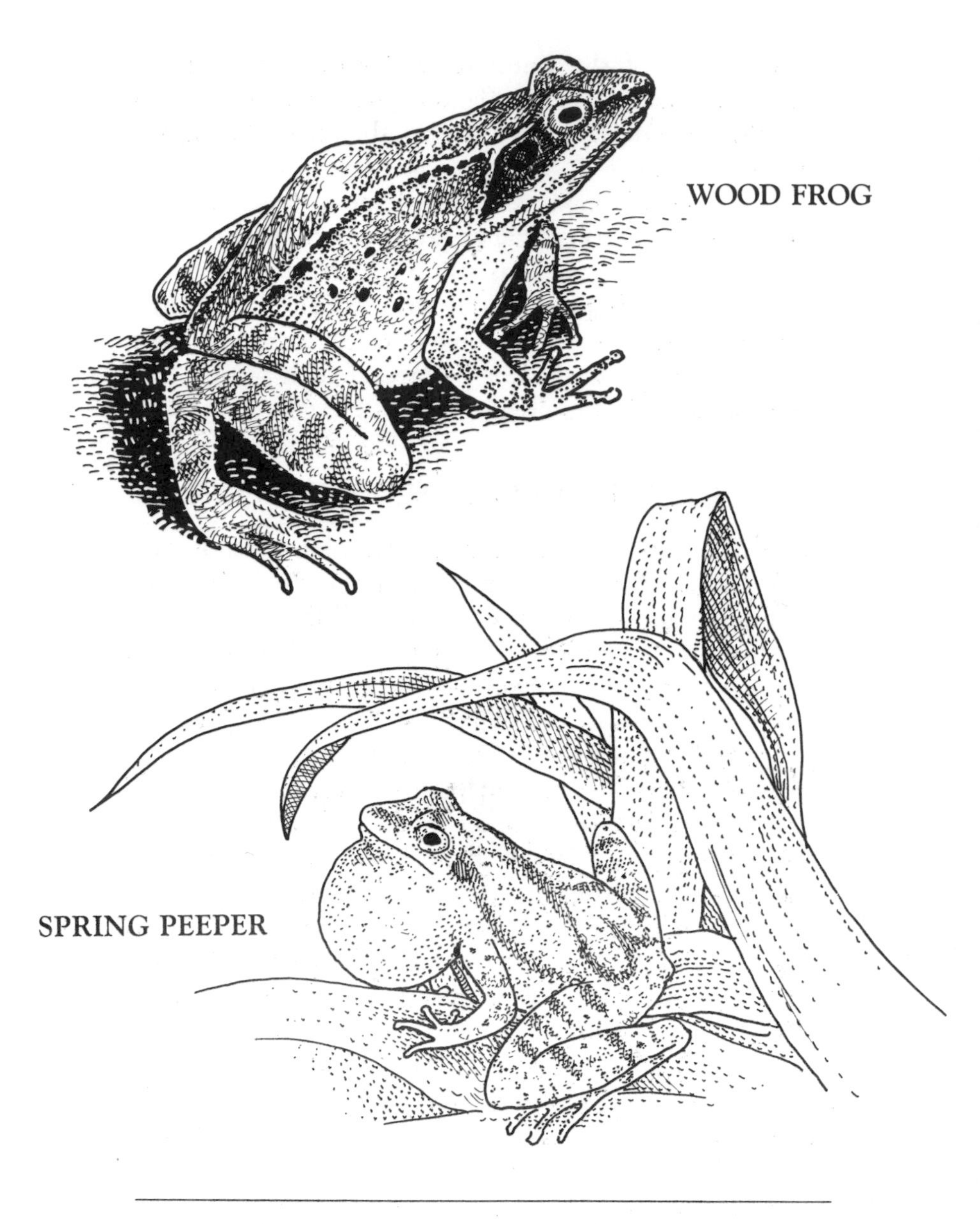

Figure 8–8. *Get to know the mating calls of different frog species and create a calendar of their occurrence.*

In many areas of the country, spring peepers are very abundant. At other seasons they are seldom seen. Their calls are heard again in fall, when day length and temperature are about the same as when they call from ponds in spring. This time their calls come from high in the trees. Little is known about where they spend the winter. There is strong suspicion that they squeeze down along the roots of trees to get below the frost. This remains to be proven conclusively. Perhaps you can be the one to unravel some of the mysteries of the day-to-day lives of these very common but largely unknown frogs.

Other frogs and toads join the frog chorus throughout the spring and into the summer. In the Northeast, toads start trilling by early May. Pickerel frogs and leopard frogs soon chime in. Green frogs and bullfrogs are among the last to join. The tadpoles of the earliest chorusers are transformed into frogs in the summer they were hatched. Green frogs and bullfrogs remain as tadpoles for a year or two.

There are records available to help you recognize the voices of the different frogs. Once you are familiar with them, begin keeping a calendar of when the first calls of each species are heard and where. Also note when the chorus for each species stops. The times will vary from year to year due to weather differences.

But don't stop with just recording the calling dates. Get out and observe the frogs. Discover the different kinds of eggs they lay. Collect small amounts of the eggs and let them hatch in an aquarium and go through their development to small frogs or toads. Then release them near where you collected the eggs. Eggs are living things that need plenty of dissolved air to grow. Even if it means taking only a part of an egg mass, take only a dozen or so eggs and keep them in the largest jar or aquarium you can to ensure their survival.

Frog numbers have been declining in many areas.

We need to learn more about why. Perhaps your investigations of local frog choruses and frog life habits will help find out why. You can help protect local frog populations by making local citizens aware of the tendency of frogs to cross roads on warm, rainy spring nights. The people can be urged to avoid running over frogs with cars on such nights and even to avoid traveling on some roads on such nights. In England, volunteers get out and help move toads across highways to breeding ponds. They collect them in buckets on one side of the road and carry them over to the other side. They even set up Toad Crossing signs to alert motorists.

For Further Reading

IDENTIFICATION GUIDES

To identify the plants and animals you find you will want to refer to field guides. There are many on the market and they cover a broad range of living things. Rather than list all the titles, only the series and the publisher are given below, with an occasional list following of the volumes that are particularly useful.

The Golden Nature Guide series, published by Golden Press. *(Weeds, Pond Life, Spiders, Seashores, Nonflowering Plants)*

The Peterson Field Guide series, published by Houghton Mifflin. *(Birds, Animal Tracks, Ferns, Wildflowers of Various Regions, Bird's Nests, Fish, Eastern Forests, Seashores)*

The Audubon Society Field Guide series, published by Alfred A. Knopf. *(Mammals, Butterflies, Mushrooms, Trees, Fish)*

The Sierra Club Naturalist's Guide series, published by Sierra Club Books. This series tells much about the habitats in different regions of the United States. *(Southern New England, The Piedmont, The North Atlantic Coast, The North Woods, The Sierra Nevada, The Deserts of the Southwest)*

An excellent guide to all the birds of North America is: *Field Guide to the Birds of North America* by the National Geographic Society, 1983.

An excellent illustrated guide to a wide range of plants and animals is: *North American Wildlife* by Reader's Digest, 1982.

GENERAL BOOKS ON DISCOVERING NATURE FOR YOURSELF

Brown, Tom, and Brandt Morgan. *Tom Brown's Guide to Nature Observation and Tracking.* New York: Berkley Books, 1983.

Brown, Vinson. *The Amateur Naturalist's Handbook.* Englewood Cliffs, N.J.: Prentice-Hall, 1980.

Fadala, Sam. *Basic Projects in Wildlife Watching.* Harrisburg, Penn.: Stackpole Books, 1989.

Leslie, Clare Walker. *Nature Drawing.* Englewood Cliffs, N.J.: Prentice-Hall, 1980.

Roth, Charles E. *The Wildlife Observer's Guidebook.* Englewood Cliffs, N.J.: Prentice-Hall, 1982.

Roth, Charles E. *The Plant Observer's Guidebook.* Englewood Cliffs, N.J.: Prentice-Hall, 1984.

GENERAL BOOKS ON ASPECTS OF OBSERVING WILDLIFE

Brewer, Jo, and Dave Winter. *Butterflies and Moths.* New York: Prentice-Hall Press, 1986.

Herberman, Ethan. *The City Kid's Field Guide.* New York: Simon & Schuster Books for Young Readers, 1989.

MacFarlane, Ruth B. *Making Your Own Nature Museum.* New York: Franklin Watts, 1989.

Nero, Robert W. *Redwings.* Washington, D.C.: Smithsonian Institution Press, 1984.

Pyle, Robert M. *Handbook for Butterfly Watchers.* New York: Charles Scribner's Sons. 1984.

Stokes, Donald W. *A Guide to Nature in Winter.* Boston: Little, Brown, & Company, 1976.
Stokes, Donald W. *A Guide to Bird Behavior, Volume I.* Boston: Little Brown & Company, 1979.
Stokes, Donald W. *A Guide to Observing Insect Lives.* Boston: Little Brown & Company, 1983.
Stokes, Donald, and Lilian Stokes. *A Guide to Enjoying Wildflowers.* Boston: Little Brown & Company, 1985.
Stokes, Donald, and Lilian Stokes. *A Guide to Animal Tracking and Behavior.* Boston: Little Brown & Company, 1986.
Tyning, Thomas F. *A Guide to Amphibians and Reptiles.* Boston: Little Brown & Company, 1990.
Vessel, Matthew F., and Herbert H. Wong. *Natural History of Vacant Lots.* Berkeley: University of California Press, 1987.

Index

About the Author

Charles E. (Chuck) Roth is a career environmental educator and author. He holds a master's degree from Cornell University in conservation education. Mr. Roth worked for many years for the Massachusetts Audubon Society, where he was involved in numerous nature education activities and served as publisher of the society's magazine for beginning naturalists of all ages entitled *The Curious Naturalist.* He has written many articles and books for both children and adults. Chuck Roth lives in Massachusetts.